IEE TELECOMMUNICATIONS SERIES 7

SERIES EDITORS: PROFESSOR J.E. FLOOD AND C.J. HUGHES

Elements of telecommunications economics

Elements of telecommunications economics

S. C. LITTLECHILD

Professor of Commerce
Head of the Department of Industrial Economics and Business Studies
The University of Birmingham
Birmingham, England

PETER PEREGRINUS LTD.
on behalf of the Institution of Electrical Engineers

Published by The Institution of Electrical Engineers, London, and New York
Peter Peregrinus Ltd., Stevenage, UK, and New York

British Library Cataloguing in Publication Data

Littlechild, S C
Elements of telecommunications economics. -
(Institution of Electrical Engineers.
Telecommunications series; 7).
1.   Telecommunication systems - Economic aspects
I.   Title II. Series
338.4'7'62138   HE7631   79-41178

ISBN 0-906048-17-6

Printed in England by A. Wheaton & Co., Ltd., Exeter

To my parents

Contents

Preface

The purpose of this book is to introduce the telecommunications manager, engineer, administrator and regulator to the whole range of work which may be described as telecommunications economics. Of course, if such a survey of the whole field is to be attempted, there is not space to treat any one topic in the depth it really deserves. For this reason, the book is entitled *Elements*. Rather than attempt to provide an exhaustive catalogue and evaluation of all work to date, I have sought to show how economists look at telecommunications, to indicate the kind of work that is being done and the reasons for it, and to exhibit the kind of results that are being obtained and some of their more important implications.

Telephone engineers all over the world will surely be familiar with *Telecommunications economics* by Mr. T.J. Morgan, now in its second edition. The present author cannot hope to rival Mr. Morgan's knowledge of telecommunications nor his experience in applying the methods of telecommunications he describes. What, then, is the case for the present book?

The justification is two-fold. In the first place, Morgan's book is entirely devoted to what is traditionally called 'engineering economics' — that is, to the appraisal of the economic merits of alternative engineering schemes. Other topics such as forecasting and accounting are, of course, covered, but only insofar as they are necessary for the development of his central theme. By contrast, engineering economics, broadly interpreted, takes up only the first seven chapters of our own book. The remainder of the book is concerned with the wide range of other topics, such as pricing and finance, cost and demand, regulation and nationalisation, efficiency and equity, with which the modern manager and administrator in telecommunications must be familiar, but which are not subsumed under the heading of engineering economics.

The second justification for the present book is that the discipline of economics itself is changing, and this is necessarily reflected in the approach to telecommunications economics. Engineering economics reflects the longstanding concern of economists with the profitable and efficient operation of the individual enterprise. However, during the last quarter century the development of 'welfare economics' stimulated economists to question whether the tariff and investment policies resulting from the application of engineering economics would lead to a use of resources which would be efficient and equitable from the point of view of society as a whole. An increasing number of economists have been concerned to trace the implications of welfare economics for telecommunications. The next five chapters of our book survey their work.

My own belief is that the character of economics will change yet again in the next quarter century. The policies implied by welfare economics have not been widely adopted, nor have they been particularly successful where they have been adopted. Economists have given too little attention to phenomena such as uncertainty and ignorance which abound in the real world, and they have failed to reconcile their prescriptions of what 'ought' to be done with explanations of what actually *is* done. In the last few years, however, there has been an increasing concern to remedy these defects. The penultimate chapter of this book discusses some of the current (and fast changing!) issues in regulation in the USA. The last chapter outlines some of the new ideas in economics and attempts to develop their implications for telecommunications.

Interest in telecommunications economics has expanded enormously during the last decade. In 1967, when I first started work on telephone pricing policy, I was not aware of more than half a dozen economists who shared this interest. Even in 1974, when I started work on this book, I still hoped to be able to reference all the published (and many unpublished) papers on telecommunications economics. Today, in 1979, it would be a full-time task merely to compile such a comprehensive bibliography, let alone to write a book. I have necessarily had to be selective in discussing material, but I have tried to reference all the published work of which I am aware. References are given in full at the end of the book, but at the end of each chapter there is a list of the works which are of particular relevance to that chapter.

My hope is that this book will be found useful in all countries of the world. If any subject is international, telecommunications is. The same problems are found universally, although different methods of solving them have been adopted. For this reason, the material in this book has been grouped by topics, illustrating these from a variety of countries. Occasionally this makes for awkward reading, but it seemed preferable to the duplication which would result if each country's telecommunications system were to be analysed in turn.

The Editor of this series, Professor J.E. Flood, made extensive and detailed comments on four successive drafts of this book, and contributed the bulk of chapter 2. He has been an invaluable source of advice and moral support. Several friends, including H. Christie, G. Pyatt, W.F. Simpson, W.M. Turner, R. Turvey and L. Waverman, kindly commented on particular chapters. I have benefited from reading early drafts of a World Bank Telecommunications Handbook, by A. Newstead, R. Saunders and J.J. Warford. Over the years there have been helpful conversations with many other people, of whom the following spring particularly to mind: W.J. Baumol, A. Charnes, W.M. Gorman, C. Horn, L.L. Johnson, M.G. Marchand, B.M. Mitchell, R. Scholnick, J. Wiseman and E.E. Zajac. Chapter 11 is based in part on research carried out jointly with J.J. Rousseau. I am grateful to those organisations which have sponsored my research or hired me as a consultant on matters of telecommunications economics, including Bell Telephone Laboratories, Illinois Bell Telephone Co., American Telephone and Telegraph Co., the International Bank for Reconstruction and Development, HM Treasury and the Economist Intelligence Unit. To all the above I offer my thanks, and hereby absolve them of responsibility.

It is a pleasure to record my gratitude to Mrs J. Tonkin, Miss K.M. Major and especially Mrs M.A. Sheridan for their beautiful typing of successive drafts of this book; to Dr. I.H. Slicer, who prepared the index; to my wife, Kate, who has suffered far too often during the writing of the book, and to my parents, to whom this book is dedicated, who have encouraged me from the very beginning.

S.C. Littlechild

Chapter 1

Introduction

1.1   The importance of telecommunications

There are at least five reasons why it is becoming necessary for more and more people to obtain a thorough understanding of the economics of telecommunications.

The first reason is the sheer expense of telecommunications systems. The Bell Telephone System accounts for approximately one-quarter of all new capital raised each year in the USA. The British Post Office invests sufficient capital to build a third London airport every six months. The World Bank lends to telephone administrations in developing countries at the rate of about one thousand dollars per minute of each working day. With investment on this scale, it is clearly important that investment appraisals be carried out efficiently, that capacity be utilised fully and that costs be reduced wherever possible.

Second, telecommunications is a complex business. There are so many interdependencies: for example, the ability to make calls between two parties in the network requires a chain of various items of equipment, each of which is being competed for by callers along many different routes. The technology is a mixture of mechanical and electronic, and the product itself is invisible! It is not easy for the layman, whether consumer, taxpayer, administrator or regulator to know whether the right decisions are being made.

Third, telecommunications is not only expanding at a rapid rate — in fact, it is one of the fastest growing industries in any country — but it is doing so, in part, because the technology is changing so rapidly. The development of coaxial cables and radio relay stations in the 1950s led to such a drop in costs and prices that demand for inland long-distance calls boomed. The new features of the 1960s were submarine cables and satellites, leading to a rapid growth of international traffic.

In the 1970s the development of electronic switching devices promises to open up an incredible variety of new services. For the future, yet other inventions such as laser beams are already well into the experimental stage.

Fourth, telecommunications systems are not merely a cost-saving device for businesses and a convenience for households. They have enormous social and political implications. If a telephone conference becomes a substitute for a meeting, how much travel will still be necessary? If news reports can be dialled on television, will newspapers become obsolete? How will family ties and the distribution of population be affected by these developments? Will the world become a 'global village'? Few of these questions as yet have answers.

Finally, we must note that, in the great majority of countries, telephone systems are run by the government. Even where they are privately owned, they are heavily regulated. Rightly or wrongly, telecommunications has been thought too important to leave to the market. Decisions about tariffs and investment are not disseminated among thousands of relatively small producers, as with food or clothing or housing. Decision making in telecommunications systems is concentrated in the hands of relatively few men, so that each decision is important for the whole country.

1.2   The nature of economics

Economics has always been concerned with *financial wealth* and matters pertaining thereto. Adam Smith's pioneering volume in 1776 was entitled *An enquiry into the nature and causes of the wealth of nations*. At the beginning of this present century it was still fashionable to describe the subject matter of economics as 'that which can be brought within the measuring rod of money' (Pigou, 1920). Following this line of thought, one would deduce that telecommunications economics is concerned with those aspects of telecommunications systems where money is involved — that is, with tariffs, investment appraisals, operating employment policies and the raising of capital.

This definition causes certain problems. A wife performing housework is not part of economics, but a cleaning lady paid to do the same tasks is! Whether the performance of an orchestra should be classed as economic depends on whether the musicians are paid! It came to be argued instead that economics should be concerned with the *allocation of resources*, whether or not monetary payment was involved (Robbins, 1932). A housewife has to allocate her weekly income among the goods in the supermarket, but she also has to allocate her time (another scarce resource) between her family, her job, her housework, her leisure etc. Different ways of spending time and money give different satisfactions. It is never possible to obtain at once all the satisfactions one would like. A choice has to be made. Economics is, therefore, often called 'the science of choice'. *Economising* is the efficient allocation of scarce resources among multiple competing ends.

This definition is obviously broader than the previous one. It means that social activities such as churchgoing and marriage have economic *aspects*, hence the study of economics may be able to throw some light on such phenomena. All those phenomena listed in the previous paragraph still belong within the scope of telecommunications economics, for they are all the subject of decisions. However, the new definition would include, for example, the allocation of frequencies in the radio spectrum between competing users, even if no price were charged.

Several economists are now moving one stage further, and asking such questions as — Who takes these decisions? With what aims in mind? How is the list of available alternatives determined? What influence does the structure of institutions in any country have upon the way in which these decisions are taken?

Attempting to answer these questions makes it abundantly clear that in any telecommunications system there is not one but many decision makers. They will generally have different knowledge, different aims and different incentives. For example, decisions about provisioning of plant may be taken by telephone engineers concerned not to overinstall expensive capacity, but at the same time concerned lest they are taken by surprise if demand suddenly rises. Capacity is expensive because other firms wish to use the necessary steel and labour for their own purposes. Demand is determined by firms and households deciding whether to conduct their business by telephone or letter. The government may be breathing down the necks of the telephone managers because it begrudges them capital, but at the same time fears the political implications of starving them.

Thus telecommunications economics is inevitably concerned not only with the decisions of telephone managers, but also with the decisions of customers, competitors and suppliers; employees, shareholders and taxpayers; regulators and administrators, politicians and governments. A comprehensive treatment of telecommunications economics will have to embrace not only the telephone administration itself, but also the social, legal and organisational environment within which it is located.

Some economists would argue that economics should be defined more widely as 'the science of human action' (Mises, 1949). The notion of economising as defined above enables one to understand how a particular decision is made, given the set of ends and given the resources available. But it does not explain where those ends or resources come from, or how they change over time. What economics must also encompass, according to these writers, are the notions of *imagination* and *alertness to new possibilities* (Shackle, 1972, Kirzner, 1973). Without these elements of change, decisions could be made once and for all, and new situations would not be recognised, nor would any improvement in satisfaction be possible.

For the most part, economics generally, and telecommunications economics in particular, is based on the notions of economising and the efficient allocation of resources, but we shall see later that the notions of imagination and alertness have a key role to play at several points.

1.3   Economics and telecommunications

A study of economics can offer understanding — an understanding of what has been, what is and what will be. This branch of economics is generally referred to as *positive* economics; it stands in contrast to *normative* economics which deals with the prescription of what *ought* to be done.

The positive economics of telecommunications should be able to shed light on questions such as the following: Why has telephone penetration grown so fast this century? Why do some countries have thirty telephones per hundred population, and others only three? What will be the level of penetration in Britain in ten years' time? Why have US telephone companies previously favoured flat-rate service for local calling areas, but are now turning to 'usage-sensitive' pricing? Who will gain and who will lose by this? Why were telephone companies nationalised in so many countries? What explains the increasing competition to enter the telecommunications market in the United States?

The answers to such questions are obviously necessary as a basis for normative economics, i.e. if one is to recommend policies for the government or the telephone administration to follow. Should prices of peak-hour calls be raised? Should investment be cut back? How fast should electronic switching be introduced? Should the telecommunications sector of the British Post Office be split off from its postal sector? To answer such questions, one needs to specify the aims that are to be achieved, the alternatives available and the consequences of choosing each alternative. It is necessary to understand how the whole economic system is likely to work, but this alone is not sufficient to determine the optimal policy. One must also make judgments about what is desired and what is not desired, one must weigh benefits here against benefits there, balance losses to one group against losses to another. For example, should the burden of financing research be placed on tax-payers, on current consumers or on future consumers? Do the advantages of avoiding unemployment outweigh the disadvantages of bringing forward the commissioning of plant not yet required? These are the kind of issues with which one must grapple. In the last resort, economics cannot dictate what should be done, but it *can* clarify the consequences of doing anything, and that may be a crucial contribution.

1.4   The literature on telecommunications economics

The earliest treatment appears to be a book by Lee (1913) entitled *Economics of telegraphs and telephones.* Famous telephone engineers like Erlang and Moe were, of course, concerned with the economic merits of alternative engineering schemes relating to the provisioning of plant, the design of exchanges, the replacement of equipment, the routing of traffic, overall system planning etc. (Syski, 1960, p. 649). Here the task is to design an engineering scheme which minimises total costs, subject to giving satisfactory performance. Such cost studies, along with methods of forecasting demand and interpreting accounts, are generally called *engineering economics* (Grant, 1938). Sir Frank Gill (1943) devoted his presidential address to the IEE to this very topic. As mentioned in the Preface, an excellent and up-to-date exposition of engineering economics applied to telecommunications is provided by Morgan (1976). No doubt material of a similar kind has been developed for use within many telecommunications entities.

Perhaps because telecommunications systems have grown so fast, and because their complexity has daunted the uninitiated, economists themselves wrote little about the economics of these systems before the 1960s. However, during the last decade there has been a growth of research and writing on two areas of telecommunications economics which may even surpass the rate of growth of telecommunications itself! Econometric techniques have been extensively applied to forecasting demand, and most telecommunications entities now have a statistical forecasting unit. Economists at universities, at Bell Telephone Laboratories and in government departments have attempted to deduce the optimal telephone pricing and investment policies implied by welfare economics; the early paper by Hazlewood (1951) is still the clearest exposition. Most of the published work has been of a qualitative kind, but increasing numbers of quantitative studies are appearing.

As a result of this work, much more is known about all aspects of telecommunications economics than anyone could possibly have expected even a decade ago. However, the answers to all our questions are by no means fully known. Occasionally the existing answers change, and certainly new questions arise. It is a salutary, as well as rewarding, experience to read the account and analysis of early telephone systems provided by a master such as Kingsbury (1915). His grasp of economics, technology and administration of telephone systems across the world is unlikely ever to be matched.

1.5   Outline of the book

The 14 chapters of this book fall into five groups. The first two chapters are introductory, designed to bring a wide spectrum of readers to a common starting point. The present chapter has sought to explain to engineers, managers and administrators the nature of telecommunications economics. The next chapter outlines for economists and other nonspecialists the elements of telecommunications technology. Throughout the book no previous knowledge of economics or engineering is assumed.

Chapters 3 to 6 describe, and illustrate with a case study, the basic economic material of any telecommunications system, namely, the demand, production and cost functions.

Chapters 7 and 8 outline various techniques for choosing investments and designing tariffs. The assumed objective, as usual in engineering economics, is either to minimise cost or to maximise profit.

The next four chapters explain the origins and nature of welfare economics, and trace the implications for pricing and output of a telecommunications entity assumed to accept these principles. There is a brief discussion of the present administration of telecommunications in Great Britain from the point of view of these principles.

Chapter 13 discusses some of the current issues in regulation in the USA. The final chapter outlines some of the newer ideas of economists on property rights and information, and suggests some implications for the administration of telecommunications.

Chapter 2

Elements of telecommunications

2.1 General

This chapter is intended to provide an elementary introduction to telecommunication services and the systems used. It describes the technical features of systems, which determine their applications and their economics, and explains some of the most important technical terms used. It is primarily written for nonengineers, such as economists, and may safely be omitted by the telecommunication engineers and managers who read this book.

The aggregate of the telecommunications plant in a particular country constitutes its *national network* (Flood, 1975). The national networks of the different countries of the world are interconnected by an *international network*, having component parts in every country and across each ocean, to provide communications between these countries. Today, this network links over 300 million telephones (AT&T, 1978). Moreover, it is continually growing, doubling its size every 10 years or so.

The ubiquitous nature of telecommunications, together with its growth, places two dominating requirements on the design of all the systems used. The first is standardisation; every new system is required to interwork satisfactorily with every obsolescent system still present in the network. The second is the ability to expand; any system may require to be extended to several times its original capacity during its operating life.

2.2 Telegraph and data systems

The electric telegraph was one of the earliest uses of electricity, and successful telegraphs were demonstrated by Wheatstone in the UK and Morse in the USA in 1837. In the Morse system, a message was sent by the operator making and breaking the current in the line to provide a sequence of short and long pulses (dots and dashes). These generated sound signals at the receiving station which were decoded by the operator there.

Nowadays, the Morse code has largely been replaced by the use of *teleprinters*, or *teletypewriters*. The operator at the sending station types the message on a keyboard, which causes it to be sent over the line in machine code (Smith, 1974). At the receiving station, the message is printed by the machine without requiring an operator. Messages may also be sent from and received on *punched tape*.

In the public telegraph system, messages are handed into the telegraph office by customers and are delivered to recipients from the receiving office by messengers. The system is of diminishing importance because of falling traffic, owing both to the high cost of employing messengers and the greater convenience to most customers of using the telephone.

Unlike the public telegraph system, the *telex* system, in which messages are transmitted directly between teleprinters on subscribers' own premises, is growing every year (Ronton, 1974). This system requires each subscriber's teleprinter to be connected to a switching centre or exchange so that he can 'dial up' the connections required. The telex system has advantages over the telephone system when 'hard copy' of messages is required. Consequently, telex is used mainly for business purposes. Another advantage is that messages are received without requiring anybody to be present at the time. This is convenient for communication between countries in different continents, which have large time differences. For example, a message may be sent from the UK during office hours, received in Australia during the night and read by the recipient on the followihg morning. As a result, a high proportion of the traffic handled by the telex system, unlike the telephone system, consists of international traffic.

In the telex system, a circuit between the sending and receiving stations must be maintained by the switching centres concerned for the length of time required to transmit a message. This mode of operation of the switching centres is known as *circuit* switching. An alternative mode of switching is called *message switching* (Robin and Hallo, 1966). In this second method, the circuits between switching centres are never directly connected together. A message arriving at one switching centre and destined for another is stored at the first centre until a line to the other becomes free and is then sent over it. In this way, the lines can be occupied for 100% of the time during the busy part of the day. This is not possible with circuit switching, so message switching produces savings in the number of circuits required at the cost of greater complexity in the switching centres. Delay is incurred while messages are queuing for circuits to become free. This prevents the use of message switching for telephones, but it is not a disadvantage for telegraphy. Message switching used to be done by storing telegraph messages on paper tape and tearing the tape off the receiving machine for onwards transmission by another machine; this was known as *torn-tape* operation. Nowadays, the storage is usually electronic, the switching centres being controlled by computers (Flood, 1975, Rubin and Hallo, 1966).

Message switching is often used for international telegraphy. The high cost of the long international circuits used makes it obviously desirable to use them as efficiently as possible. Message switching is also used in military applications and in the private telegraph networks of large organisations such as airlines. These organisations have an incentive to minimise the number of lines which they lease from the public telecommunication administrations for their private use.

Nowadays, messages are also transmitted between computers and between remote terminals and computers. This is known as *data transmission*. It usually requires a higher speed of transmission than is provided by telegraphy, so the messages are transmitted over telephone lines by using equipments known as *modems* (modulator – demodulators). Sometimes these messages are transmitted over dialled-up connections through the public switched telephone network. However, a user's volume of traffic often justifies the leasing of private circuits. Some large organisations lease sufficient of these circuits to constitute a private data network. These networks sometimes use a form of message switching known as *packet switching* (Davies and Barber, 1973). In packet switching, a long message is broken down into small sections, called packets, which are transmitted separately, thereby avoiding long delays to other messages.

2.3   Telephony

The first practical telephone was invented by Alexander Graham Bell in 1876. His idea was to 'make a current of electricity vary in intensity precisely as the air varies in density during the production of sound'. To do this, he developed a transmitter (or microphone) to convert the speech vibrations into an electrical signal and a receiver to convert this signal back into a reasonable copy of the original speech. By connecting the transmitter and receiver to a line, conversation became possible between people separated by a considerable distance. The obvious advantages of immediate 2-way communication between users, without the intermediary of an operator, led to the growth of today's telephone service and the stultification of the public telegraph service.

2.4   The network

Telephone users are usually called *customers* or *subscribers*. The location of the customers telephone is called a *telephone station*. When there are a large number of telephones, from which any customer may wish to speak to any other, it is obviously uneconomic to provide a line from each telephone to every other one. Consequently, each subscriber's station is connected by a *subscriber line* to a switching centre, which connects together an appropriate pair of subscribers' lines, as required, for each telephone call. The switching centre is known in the UK as a *telephone exchange* and in North America as a *central office*. The network of lines connecting the telephone stations to the switching centre is called the *local network* or the *subscribers' distribution network*. Each path which is set up temporarily through the network for a telephone call is called a *connection*.

If a large town contained only a single switching centre, many of the lines would be very long and the cost of the network would be excessive. It is then more economic, as shown in Fig. 2.1, to divide the town into separate areas, each served by its own switching centre. This is because the cost of providing additional switching centres is more than offset by the reduced cost of the shorter cables required. However, customers whose telephones are connected to different switching centres will wish to converse together. It is therefore necessary to provide lines between the different switching centres, so that calls can be made over connections involving two switching centres in tandom. A line interconnecting two switching centres is called a *junction* in the UK or a *trunk* in the USA. The network comprising all these lines is called the *junction network*.

In a large city area, there may be a very large number of switching centres. It therefore becomes uneconomic to provide junctions between all the switching centres (just as it is uneconomic to provide lines between all telephones in a local area). Instead, an additional switching centre is installed, which has junctions to all the other switching centres and serves solely to make connections between them. This switching centre is called a *tandem exchange* or *tandem office*. The

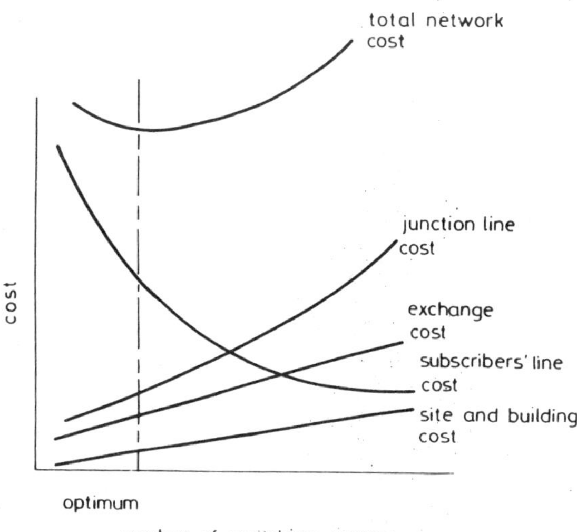

Fig. 2.1   *Variation of network cost with number of switching centres*

switching centres connected to subscribers' lines, to distinguish them from the tandem centre, are called *local exchanges* in the UK and in North America they are called *end offices* or *class-5 offices*.

As the performance of telephone instruments and line plant improved, it became possible to make intelligible conversations over long distances. This led to the towns and cities of each country being joined by long-distance circuits. These are known as *trunk circuits* in the UK and *toll circuits* in North America. They comprise the *trunk network* or *toll network*. The switching centres which they link together are called *trunk exchanges* and *toll offices*, respectively. The major switching centre in each town, which connects its junction network to the toll or trunk network, is known internationally as the *primary centre*. However, in the UK it is called a *group switching centre* (GSC) and in North America as a *class-4 office*.

It is usually uneconomic to provide direct circuits between every primary centre in a country, particularly if it covers a large geographical area. Consequently, the trunk network contains switching centres interconnecting trunk circuits to enable calls to be made between primary centres which are not linked by direct circuits. These are called trunk or toll *transit centres*. Internationally, these are termed secondary and tertiary centres. In North America they are known as primary, sectional and regional centres (classes 3, 2 and 1) and in the UK as district and main switching centres.

The national *public switched telephone network* (PSTN), as shown in Fig. 2.2, can be seen to consist of a hierarchy of networks, each with its own switching centres. This enables calls to be made, as desired, between any pair of telephones in the country. Whenever a route does not exist at a particular level in the hierarchy, the connection can be made by routing it through switching centres at a higher level. Often more than one path through the PSTN exists between the same two places, and if one is busy, another can be used; this is known as *alternative routing*. Thus calls between the north and south of the USA can be switched from the East Coast during the morning traffic peak to the West Coast, where there is usually spare capacity because of the time difference. Finally, the national network is connected by an international *gateway* exchange to the international network which links the countries of world together.

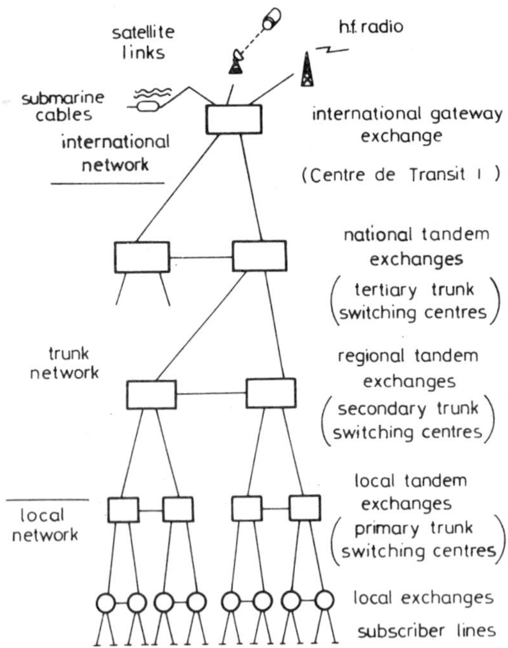

Fig. 2.2   *National telecommunication network*

The telephone network can be seen to consist of three different kinds of plant (Hamsho, 1967, Flowers, 1976): *subscribers' apparatus* (located at telephone stations), *switching equipment* (located at the switching centres) and *outside plant* (between subscribers' stations and switching centres and between switching centres). To provide satisfactory speech transmission over long distances the outside plant needs to be supplemented by electronic equipment such as ampliers; these constitute *transmission systems* (Bell Telephone Labs, 1970, Hills and Evans, 1973). To establish the required connections between subscribers' stations (and to disconnect them when calls terminate), it is necessary to send signals over the network (both between subscribers and switching centres and between different switching centres). This is done by *signalling systems* (Welch, 1979), whose apparatus has to be incorporated in both switching and transmission equipment. An introduction to the principal features of these different kinds of plant will be given in the following Sections.

2.5   Subscribers' equipment

The essential apparatus at a subscribers' station is a *telephone set*. This consists mainly of the handset (containing the transmitter and receiver), the dial and the bell. In addition to the basic telephone set, it is possible to install a number of optional extras (called *vertical services* in the USA and *premium services* in the UK). These include more attractively styled telephones, longer telephone cords, extension telephones, push-buttons instead of dials (known as keyphone in the UK and touchtone in the USA), loudspeaking ('hands-free') telephones etc.

The larger customers also have their own switching systems to enable calls to be made between their extension telephones and between extensions and the PSTN. These may either be manual switchboards, known as private manual branch exchanges (PMBXs) or automatic systems, known as private automatic branch exchanges (PABXs). Some customers use an internal telephone system which is entirely separate from the PSTN and in which connections between telephones are made by a private automatic exchange (PAX). In some cases, customers may link their PAXs in different locations by private lines to provide a private telephone network for their own exclusive use. This is economic if the amount of traffic between the different locations justifies the cost of leasing lines from the operating administration instead of paying for calls over the PSTN.

2.6   Switching systems

Originally, telephone exchanges used manual switchboards, with an operator connecting lines together as requested by customers. The first practical automatic switching systems was invented in 1889 by

Almon B. Strowger, an undertaker in Kansas City who allegedly found that his calls were being diverted to a rival by one of the local telephone operators. He therefore attempted to devise a 'girl-less, cuss-less telephone system'. The *Strowger system* was in use by 1912 and for almost 50 years it was the standard system in most countries of the world. Although it has largely been superseded by more modern systems for current manufacture, the majority of telephones in the world are still served by the Strowger system.

Strowger's invention replaced the operator searching over the switchboard for the appropriate jack into which to plug a cord by an *electromechanical selector* moving metal fingers (known as wipers) over a bank of contacts to reach the required outlet. The subscriber's action of requesting a number from the operator was replaced by means of a dial on the telephone. The Strowger system operates on the *step-by-step* principle (Smith, 1974). Each dialled digit directly operates a successive stage of selection.

The step-by-step system has a number of advantages, which accounted for its pre-eminence for so many years. Its concept is simple and it is relatively cheap to manufacture and straightforward to maintain. Because each selector is an independent mechanism, failures can only affect connections whose paths contain the faulty selectors. Thus no single fault causes an appreciable deterioration in the quality of service provided. However, the disadvantages of the system have led to its ultimate eclipse. Its complex mechanisms require considerable attention by maintenance staff and to this increases costs when wages rise. The rigid association between numbers dialled and routes selected became an increasing disadvantage as networks grew and the routings required increased in complexity. Finally, the services which switching systems are required to provide for their customers, known as *facilities*, have increased. This has resulted in additional equipment being grafted onto the Strowger system in ways that are often uneconomic.

Flexibility of routing is an essential requirement of the modern PSTN; it requires the facility for numbers used to set up interexchange connections to be different from the numbers dialled by subscribers and to be readily changed when the configuration of the network is altered to accommodate growth. The need first arose with the development of the junction networks of large cities; it is obviously undesirable for calling subscribers in different parts of the city to have to dial different routing digits to reach the same called subscriber. The need became paramount when subscribers were enabled to dial their own long-distance calls, known in the USA as *direct distance dialling* (DDD) and in the UK as *subscriber trunk dialling* (STD). The provision of this facility requires that each subscriber be allocated a single *national number* which can be used to reach him regardless of the place where the call originates and routing of the required connection. Finally, subscribers have been enabled to dial their own international calls, known as *international subscriber dialling* (ISD). This necessitates an *international numbering scheme*, whereby a call can be made to a given country by dialling the same code digits from any other country in the world.

Flexibility of routing is provided by the facility known as *translation.* It requires the switching system to have equipment known variously as *registers, register–translators* and *senders*. The digits dialled by a subscriber are received by a register and stored until sufficient digits have been accumulated to determine the required routing. The routing digits dialled by the subscriber are then translated, i.e. replaced by a different set of digits which are sent out from the register to set up the connection. Since a register is required only during the short time while a connection is being set up, and not during the subsequent conversation, relatively few registers are required to handle the traffic. They can be arranged in a common pool and connected to the switches only when they are required to receive digits and control the setting up of connections. Equipment used in this way is called *common-control equipment.*

Registers have been successfully added to the Strowger system, first during the 1920s to cater for the needs of large cities (the director system) and later to provide for long-distance calls to be dialled (STD registers installed in GSCs). However, common control is an inherent feature of more modern switching systems. The most popular example of these is the *crossbar system* (Rubin and Haller, 1966, Flowers, 1976).

The crossbar switch was invented by Palmgren and Betulander in 1919, but the system was not widely used until the 1950s. The crossbar switch consists of a rectangular array of contacts, operated by electromagnets through horizontal and vertical bars. Because of its inherently simpler mechanism and because it is able to use precious-metal contacts, the crossbar switch requires much less maintenance attention than the Strowger switch. However, its mechanism cannot be directly controlled by the subscribers dial. It is necessary for the dialled digits to be received by a register and for operating signals to be applied to the switch magnets by another common-control equipment called a *marker.* Since a marker requires only a few seconds to operate a switch, only a few markers are required in even the largest crossbar exchange. Consequently, a faulty marker can result in the failure of a large number of calls. Elaborate measures are therefore taken to minimise the occurrence of faults and to detect and record them when they do happen. If the number of calls offered to the markers is greater than they can handle, the effect is similar to switches being out of service. The crossbar system thus suffers much more than the step-by-step system from the effects both of common-control faults and overload by unexpected peaks of traffic. For many years, the need to use common control placed crossbar at an economic disadvantage compared with Strowger. However, when registers became essential to provide network flexibility this disadvantage disappeared and the reduced maintenance costs of crossbar switches led to the widespread adoption of the system.

During the last two decades there has been considerable interest in *semielectronic* and *electronic switching systems*. Fully electronic systems have recently been developed for private exchanges (IEE, 1978), but the public-exchange systems which have been developed are mainly semielectronic (Joel, 1976). These use either miniature crossbar switches, or replace them by arrays of reed relays. The latter not only use precious-metal contacts, but are completely sealed and thus provide even greater reliability than the crossbar switch. In the common-control equipment of these systems, electromechanical markers (as used in crossbar) have been replaced by an electronic *central processor*, which is a special-purpose electronic computer. Systems of this kind have been developed in the USA (e.g. the ESS no. 1 system), in the UK (TXE2 and TXE4), in Europe and in Japan, and they are now being installed in considerable numbers.

As a result of their computer control, electronic switching systems can easily provide new facilities for subscribers, such as automatic call transfer to temporary numbers, conference calls, abbreviated dialling, etc. They also provide additional facilities for the operating administrations, such as the ability to change subscribers' numbers and the range of facilities provided (known as the subscribers' *class of service*) without physical changes to the wiring of the exchange. Although a single central processor can operate sufficiently fast to handle all the calls of the largest exchange, it must be at least duplicated to provide *security*, i.e. continuity of service in the event of faults. The measures required to detect faults and ensure that faulty equipment is removed from service must be more comprehensive than in a crossbar system because the possibility of a fault causing a complete breakdown is even greater.

The first phase of automatic telephony, the introduction of electromechanical systems, reduced the need for operators, but increased the need for engineering maintenance staff. It provided subscribers with a service which was faster, but less flexible, than that given by operators. The latest phase, the introduction of electronic systems is restoring flexibility by providing automatically facilities previously requiring an operator. It will also reduce the need for engineering staff, thus enabling the network to continue its growth without creating excessive manpower requirements.

2.7 Billing

One facility which is required of all switching systems is a means of recording charges for calls to enable operating administrations to recover the costs. There are two basic methods: *metering* and *ticketing*. Both methods were used with manual switchboards. An operator charged for a call either by pressing a key to operate a counter associated with the subscriber's line or by writing down the details of the call on a ticket. The former method was used for local calls costing only a single unit fee and the latter for long-distance calls for which the fee must be calculated from distance and duration. Both methods are still employed with automatic switching systems.

When automatic systems were first introduced they provided only local connections. Some operating administrations, particularly those in Europe, introduced *message rate* charging. The switching equipment operated the subscribers' meter once for each local call made. When long-distance dialling was introduced, the equipment was modified to provide *periodic pulse metering* (PPM). The meter is operated at intervals throughout the call and the duration of these intervals is determined by the destination of the call, thus resulting in a charge proportional to time and distance. Other administrations, particularly in the USA, introduced flat-rate charging. No charge was made for local calls, the rental being fixed to cover the cost of calls made by the average subscriber (either residential or business). Since subscribers' lines in flat-rate areas are not equipped with meters, it was not convenient to use PPM when long-distance dialling was introduced. Instead, equipment was added to provide automatic toll ticketing (ATT) or automatic message accounting (AMA). This records, usually on punched tape, the same details previously recorded by an operator and the charge for the call is calculated subsequently on a computer.

Automatic ticketing enables the subscribers to receive an *itemised bill* giving full details of each toll call made. However, because of the cost of the equipment and the data processing, a fairly large minimum charge is retained, usually that for a 3 minute call. When metering is used, the subscriber can be provided only with a bulk bill giving the total cost of all calls. However, subscribers can make short-duration calls even over long distances at low cost, the minimum charge being a single unit. This largely removes the demand for special services requiring operators, which continue to be required when ticketing is used. For example, instead of waiting for an operator to connect a person-to-person call, the caller can make a short call (costing one unit fee) to the called subscriber's usual number to enquire whether he is there.

2.8   Congestion

Switching systems are not dimensioned to cater for every subscriber being engaged in a call simultaneously, because this is most unlikely to happen. Thus, congestion sometimes occurs, i.e. call attempts are unsuccessful because all the equipment suitable for use in the required connections is already busy. In practice, sufficient equipment is provided to carry the traffic occurring during the busiest period in the day (known as the *busy hour*) with a probability that calls will be unsuccessful due to congestion that is considered to be sufficiently small to be accepted without complaint. This probability is called the *grade of service*. Thus, a high value of grade of service corresponds to poor service to subscribers but provides economy in equipment, and vice versa. Values that are used in practice vary from a few lost calls in every thousand for connections within an exchange (which are relatively cheap) to one in ten for very expensive long-distance circuits.

To dimension the equipment, it is necessary for the engineer to know both the grade of service which is specified and the amount of *traffic* to be carried. The measure adopted for the latter is simply the average number of calls in progress during the busy hour. The unit in which it is measured is called the *erlang* (E), after A.K. Erlang, the Danish mathematician who established the basic theory of telephone traffic in the early 1900s. Since a large number of short-duration calls produce the same traffic as a small number of lengthy calls, the traffic may be expressed as

$$A = nh/T \tag{2.1}$$

where A is the traffic in erlangs, n is the number of calls originating in time T and h is the mean holding time (i.e. average duration) of the calls. In the USA, traffic is often measured in hundreds of call-seconds (CCS) per hour instead of in erlangs. From eqn. 2.1, $1 \cdot 0$ E = 36 CCS per hour.

It will be noted that the cost of a switching centre depends not only on the number of lines but also on their *calling rate* (i.e. the amount of traffic they originate). The latter may vary between 0·01 E for residential subscribers and 0·4 E for heavily used business lines. In addition, there is a cost element which is almost independent of the number of lines. This arises from the costs of land, building, power plant and (for electronic systems) central processors.

2.9   Outside plant

Originally, open-wire lines on poles were used. These have now largely been replaced by cables, varying in size from two pairs to 2000 pairs, which may either be overhead or underground. In urban areas, underground cables are laid in ducts; this enables additional cables to be added to a route without the inconvenience and cost of re-excavation. In local networks, the cable pairs terminate at distribution points (DPs). These are convenient terminals for the connection and disconnection of subscribers' stations, as they either request or discontinue service. In rural and suburban areas, DPs are usually on poles, from which telephone stations are connected by drop wires. In city centres, they are usually underground or inside large buildings. Multipair cables are also used for junction routes and the shorter trunk or toll routes.

For long-distance routes, *coaxial cables* are used. These consist of an inner conductor surrounded by an outer conductor in the form of a hollow tube. They have the advantage that the electric field associated with the current is completely screened. This enables them to be used at high frequencies, and so carry a large number of simultaneous calls by means of carrier working, as explained in the next section. In future, long-distance cables may be composed of optical fibres conveying signals by light waves instead of coaxial tubes carrying electric currents (Martin, 1977).

Long-distance routes also use *radio links*. These operate at microwave frequencies, giving line-of-sight transmission. A prominent feature of these systems is thus the towers for the antennas of intermediate stations, spaced at intervals of about 50 km along each route. The same methods are used to provide international circuits. Coaxial *submarine cables* have now been laid across the oceans and radio links are provided by *communication satellites*.

2.10   Transmission systems

As signals travel along a line, or over a radio path, they lose their power and are said to suffer *attenuation* or *loss*. This loss is measured in decibels (dB). In a long circuit, it is necessary to install amplifiers or *repeaters* at regular intervals to provide gain to restore the loss. A transmission system therefore consists of the line (or radio path), a *terminal station* at each end and a number of *intermediate stations* containing the repeaters. Since most amplifiers amplify only in one direction, it is necessary to provide a separate transmission path, known as a *channel*, for each direction of transmission in a long-distance telephone circuit.

The loss of a line increases with frequency, and it is possible to provide only a limited amount of gain, so satisfactory transmission can be provided only over a limited band of frequencies, known as the *band-width* of the system. Clearly, this bandwidth must exceed the band of frequencies contained in the signal to be transmitted. By international agreement, the band of frequencies transmitted for a telephone channel is from 300 Hz to 3400 Hz (the units of frequency is the hertz, where 1 Hz - 1 cycle per second). Some signals, e.g. telegraph signals, require a much smaller bandwidth than telephony. Other signals require much wider bandwidths, e.g. colour television requires over 5 MHz (5 million cycles per second).

The simplest telephone transmission system is a cable pair transmitting a single channel at audio frequencies. However, when the bandwidth of a transmission path exceeds the bandwidth of the signals to be sent over it, it is possible to use it to provide several independent channels; this is known as *multiplexing*. One method commonly employed is to shift the channels to different frequency bands, as shown in Fig. 2.3; this is known as *frequency-division multiplexing* (FDM).

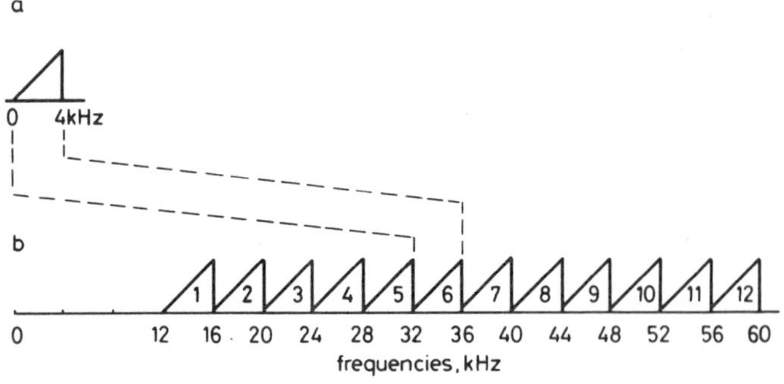

Fig. 2.3   *Frequency-division multiplexing (12-channel carrier system)*

    (a)   Frequency band of individual telephone channel (channel 6) before multiplexing

    (b)   Frequency bands of channels after multiplexing

Systems using this process are called *multichannel carrier systems*. They are employed to transmit 24 telegraph channels over a single telephone channel or many hundreds of telephone channels over a coaxial cable or a microwave radio link. In an FDM system, the signal present in each frequency slot varies proportionally to the signal of the corresponding input channel. These systems are therefore known as *analogue transmission systems*.

Another method of multiplexing the signals of a number of channels to transmit them over a wideband transmission path is called *time-division multiplexing* (TDM). Whereas FDM makes a portion of the total bandwidth available to each channel for all the time, TDM uses the whole bandwidth for each channel, but for only part of the time. In this way, samples of the signal of each channel are sent as a train of short pulses. In the intervals between the samples of one channel are sent the examples of the other channels, as shown in Fig. 2.4a.

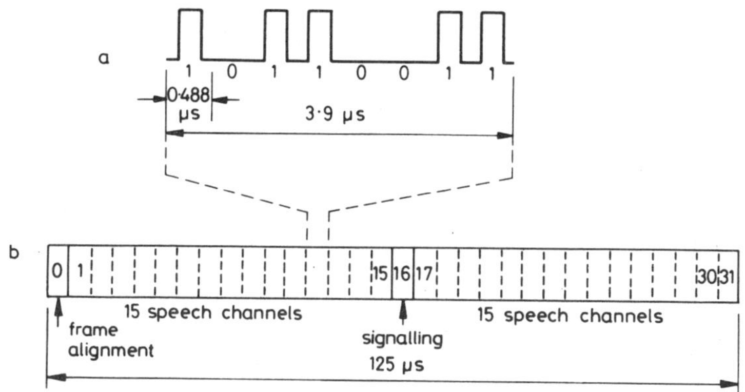

Fig. 2.4   *Time-division multiplexing using PCM*
    (*a*)   Typical digit pattern for sample of one channel (time slot 12)
    (*b*)   Frame structure of 30-channel PCM system (CCITT). Time slot 0 is used for synchronisation and time slot 16 is used for signalling

In practice, each sample is sent by means of a group of on/off pulses which represent its magnitude in binary code, as shown in Fig. 2.4*b*. This process is called *pulse-code modulation* (PCM). Systems which transmit the on/off pulses of PCM are called *digital transmission systems*. Such systems are used to transmit 24 or 30 channels over a pair of wires in a normal cable of the type designed for audio transmission. However, high-order multiplex systems are now being developed to transmit larger numbers of channels over coaxial cables and digital radio links.

A multiplex transmission system makes economic use of line plant, since many channels can be transmitted over each pair of wires. However, *channelling* equipment must be provided at the terminal stations. Thus, multiplex transmission is only cheaper than audio transmission if the cost of the channelling equipment is more than offset by the saving in cable pairs. Short-distance circuits, such as subscribers' lines and junctions between local exchanges use audio transmission, but multiplex transmission is used for long-distance circuits.

PCM requires the transmission of pulses at a rapid rate, so it requires greater bandwidth than FDM transmission. A 24-channel PCM system transmits pulses at a rate of about 1·5 MHz, whereas a 24-channel FDM system requires a bandwidth of only 96 kHz. Thus PCM makes less economic use of line plant than FDM transmission does. However, PCM channelling equipment is cheaper than FDM channelling equipment, which requires expensive filters to separate the frequency bands of the different channels. As a result, PCM is more economic than FDM for short routes (i.e. tens of kilometres), where the channelling equipment costs more than line plant and FDM is more economic for long routes, (i.e. hundreds of kilometres), where the line plant is the major component of the total cost.

Technological progress has increased the bandwidth which transmission links can provide, thus reducing the cost per channel. Over the past 25 years, the bandwidth which coaxial-cable systems can provide has increased from 4 MHz to 60 MHz and future optical-fibre systems will have bandwidths of the order of 500 MHz (Martin, 1977). As a result, it is becoming less important to make optimum use of bandwidth and this will make PCM transmission economic over longer distances. These developments should lead ultimately to *integrated digital networks*, which provide transmission and switching for a wide range of digital signals to cater for telephony, telegraphy, data transmission and many other serivces using common plant.

2.11   Ownership and control of telephone systems

As telephone technology gradually developed into a viable commercial proposition, private telephone companies were set up across the world, several under license from Bell. There was often more than one company in each country, sometimes more than one in the same town. Competition was frequently fierce, and the successful companies drove out or took over the unsuccessful ones.

In the USA, independent telephone companies flourished after the expiry of the original Bell patent in 1893. By 1907, the Bell system owned 3·1 million telephone stations while independent companies owned 3 million (Phillips, 1969, p. 655, citing Bureau of the Census 'Special Reports: Telephones, 1907', p. 22). In response to a threatened antitrust suit over its acquisition of a substantial interest in Western Union, and complaints by the independent companies that it was refusing satisfactory long-distance interconnections, the Bell System holding company American Telephone and Telegraph Co. (AT&T) entered into the so-called Kingsbury commitment of 1913, whereby it agreed to sell its Western Union stock, agreed to interconnections with those independent companies which met its equipment standards, and promised not to acquire control of competing telephone companies. The number of independent companies has nonethless decreased steadily, from over 6400 in the early 1930s to some 2000 today (Phillips, 1969, p. 650). This reduction is the result of mergers amongst the independents to achieve economies of scale. The Bell System is nowadays responsible for some 84% of the telephones in the USA, and also operates the interstate system via its Long Lines subsidiary. All the telephone companies in the USA are privately owned, but their rates are regulated by state commissions and by the Federal Communications Commission.

Apart from the United States, nowadays only a few countries such as Puerto Rico, Hong Kong and Barbados have telephone systems which are entirely privately owned, and even these are regulated by the government. In all other countries, governments have taken over part or all of the system. In Britain, for example, the Post Office was able to extend to telephones its existing monopoly over posts and telegraphs. In 1892, it bought the trunk network of the country from the National Telephone Company, and by 1912 had taken over the rest of the system, except for the municipal system in Hull. A similar pattern of events occurred in France and Belgium. In Germany, the state took over development of telephones from the very beginning, whereas in Canada, Denmark and Norway a mixed private and public network exists to this day. In Italy and Spain there exist single organisations which are jointly owned by the government and private investors.

Data assembled by the author for 114 countries in 1973 reveal that in 84 countries the government was directly responsible for telecommunications, in 19 countries there was some form of mixed or joint ownership, and in 11 countries telecommunications were supplied by one or more regulated private company. Further details of the 'early days' can be found in Kingsbury (1915). A booklet entitled *Calling the world* (1975), produced by AT & T Long Lines, contains brief summaries of telephone history provided by the countries themselves. The same organisation issues an annual booklet entitled *The world's telephones* containing statistics on number of telephones and whether under private or government operation.

In many countries, such as Britain, Austria, France and West Germany, the government operates post, telegraph and telephone services (hence the common abbreviation PTT). In other countries, such as the USA, Canada, Australia, Belgium and Sweden, posts and telephones are run by separate organisations.

2.12   Telephone tariffs

A typical telephone tariff contains three components:

(i)   An *installation charge* is paid when the telephone is first installed; this may include a refundable *deposit* or *loan*. The charge may be related to the distance betweeen the subscriber's house and the local exchange, or to the amount of installation work required. It may also vary according to the size of the local telephone network, or according to the category of subscriber (e.g. domestic or business).

(ii)   A periodic *subscription rate* or *rental* is charged on a monthly, quarterly or annual basis. This may also vary according to size of local network and type of subscriber. It will depend on the subscriber equipment selected (e.g. the number of 'optional extras') and often includes a number of 'free' local calls. It may depend on whether the exchange is automatic or manually operated.

(iii)   There is finally a *charge per call* which may depend on duration and destination of call, time of day, whether self dialled or obtained through the operator etc.

Further details of telephone tariffs in 55 countries can be found in the special issue of TELE prepared by Roos *et al.* (1976). The tariff for Costa Rica, which is typical in most respects, is set out and discussed in Chapter 6 of this book.

Chapter 3

Demand for telecommunications

3.1 Introduction

There is a great variety in the patterns of ownership and usage of telephones across different countries and over time. Ultimately, one would like to be able to explain why this is so, to ascertain how far different explanatory factors account for different patterns of behaviour. A complete explanation is not yet available, but a great deal of progress has been made over the last decade by the use of statistical and econometric techniques. This work has been done by economists in universities and by forecasters in the telephone organisations. Many telephone organisations now have very substantial forecasting units.

The material in this chapter will be presented mostly from the viewpoint of the telephone organisation wishing to explain and forecast demand, but obviously from time to time this will necessitate looking at the situation from the point of view of the customers, since it is they who make the decisions to rent or use telephones.

The first few sections of the chapter will consider the demand for connections and the later sections the demand for calls. Every textbook on economic theory contains at least one chapter on demand curves in general, e.g. Lipsey (1971), Alchian and Allen (1974) and Ferguson and Gould (1975).

3.2 Forecasting demand over time

If the telephone organisation overestimates demand for telephones, a large amount of capital will have been unnecessarily tied up for several years. If it underestimates demand, customers will have to forgo the benefits of connection. there will be waiting lists, complaints and hasty, perhaps expensive, remedial action. It is therefore important to forecast accurately (Select Committee, 1967, Chap. 7).

The most straightforward approach to forecasting is to fit a curve to the level of demand over previous years (or quarters, or months) and then predict by extrapolating this curve. Consider the linear curve

$$y = a + bt \tag{3.1}$$

where y denotes demand, t denotes the year and a and b are parameters to be estimated. A linear curve implies that demand increases by a constant amount b per year, as shown in Fig. 3.1. Such a curve has the advantage of being easy to fit using least-squares regression techniques (see any introductory statistics or econometrics text, e.g. Johnston, 1963).

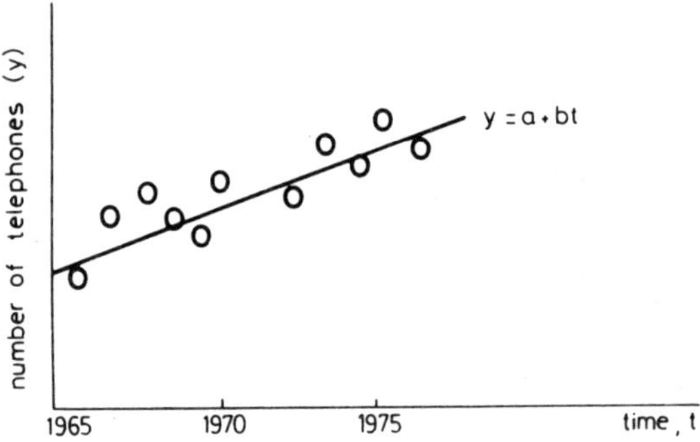

Fig. 3.1   *Demand as linear function of time*

In many situations the rate of increase in demand is itself increasing over time. Roberts and Ward (1974, Appendix 2A*i*) found that *new* demand (Δy) for telephones in Australia over the period 1962/63 to 1972/73 could be well approximated by the linear equation

$$\Delta y = 141\ 261 + 16\ 078\ t \quad (R^2 = 0 \cdot 82) \tag{3.2}$$

with $t = 0$ in 1962/63. For example, new demand in 1973/74 (year $t = 12$) was predicted as $141\ 261 + (16\ 078 \times 12) = 334\ 200$ telephones, increasing thereafter by about 16 000 telephones per year. The correlation coefficient $R^2 = 0 \cdot 82$ indicates that the constant time trend was ableto explain about 82% of the variation in new demand over the previous decade.

One can generally obtain slightly better fits by using quadratic and exponential curves instead of linear ones. These have the form $y =$

$a + bt + ct^2$ and $y = ac^{bt}$, respectively. The exponential curve implies a constant *proportional* rate of growth.

A difficulty with all these approaches is that the methods do not readily adapt to new data. The most recent observation has no more significance than the oldest one. Thus Roberts and Ward (1974, p. 7) report that 'the method [described above] has completely failed to rise to the boom in the last eighteen months for annual, quarterly or monthly forecasts'. One way of remedying this is to use exponential smoothing and autoregressive moving average (Box–Jenkins) techniques. Instead of weighting equally all previous observations, greater weight is given to observations from the recent past than to those from the more distant past. Such techniques are employed quite extensively in the British and Australian Post Offices, but the details are beyond the scope of the present book (Turner, 1973, Thompson and Tiao, 1971, Dunn *et al.*, 1971). There are numerous texts on forecasting— Wood and Filder (1976) and Lewis (1977), for example.

3.3. Saturation models

The demand for telephones cannot be expected to grow at a constant rate for ever. It has been suggested that the 'saturation' or 'epidemic' models which are widely used in biology may be applicable to telecommunications. Suppose that there is a saturation level of demand— for example, when every household has a telephone. Suppose further that the rate of increase in number of subscribers in any year is proportional to the number of outstanding nonsubscribers. The resulting level of ownership will follow an S-shaped path over time, known as a *logistic curve*.

Formally, let $y(t)$ be the average number of telephones per household at time t, and let M be the saturation level. M represents the maximum ownership level per household, which is assumed to remain constant over time. The $M - y(t)$ is the total market remaining to be exploited at time t. Assume that the proportional rate of change in $y(t)$ is a constant proportion of $M - y(t)$, so that

$$\frac{y'(t)}{y(t)} = a\,[M - y(t)]$$

where a is the constant of proportionality and the prime denotes the derivative. This differential equation has the solution

$$y(t) = \frac{M}{1 + be^{-aMt}} \qquad (3.3)$$

where b is a constant equal to $[M-y(0)]/y(0)$ and $y(0)$ is the number of telephones at time zero. This curve is illustrated in Fig. 3.2. Evidently the position and slope of the curve depend on the choice of parameters a and b.

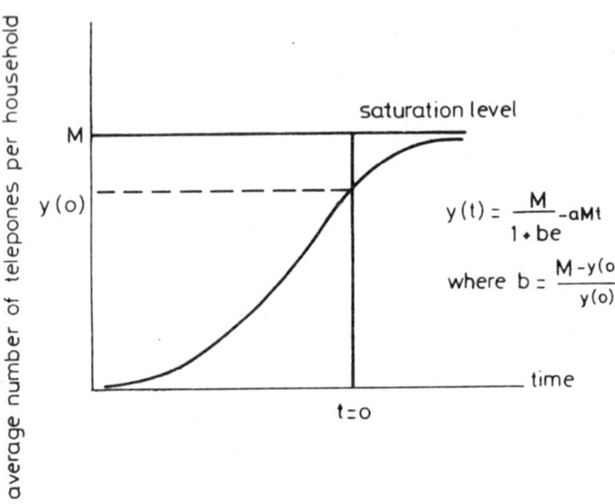

Fig. 3.2   *Demand as a logistic function of time*

The model might alternatively be developed by assuming that the number of new subscribers is proportional to the number of existing subscribers, i.e. the telephone is more attractive if there are more people to talk to. This hypothesis is discussed below (Sections 3.8 and 12.6). The *Gompertz* curve $y = ka^{bt}$ is based on similar assumptions (Flood *et al.*, 1975, chap. 13). Morgan (1976, p. 183) remarks that 'the complex mathematical manipulation required to determine the constants for this curve make it difficult to apply to the growth of telephone demand', but relatively straightforward algorithms are now available (e.g. Ullman, 1976, pp. 287–8).

One reported test of the logistic model for Bell System residence main telephones over the period 1946–69 (Chaddha and Chitgopekar, 1971) casts doubt on the hypothesis that the saturation level M remains constant over time. However, a 'generalised' model in which the saturation level changed over time as a function of disposable income provided a more satisfactory basis for forecasting. This leads up to the notion of a demand function.

3.4   The nature of demand

Forecasting techniques which relate to time alone do not explain *why* demand changed the way it did. Some further explanation would be useful. Let us consider the situation from the point of view of the customer.

A potential customer is more likely to subscribe the lower the rental charge is and the higher his income is, since in either case he is called on to make less sacrifice of other consumption goods. The potential benefits will also depend on usage, which is likely to be high when the price of making telephone calls is low and the prices of substitutes such as letters are high and when the income of the subscriber is high (or, for business customers, when the level of business activity is high).

If one could know the number of people y who would subscribe to the telephone at any given level of rental r, call price p_c, letter price p_L, income I etc., one would have what is known as a *market demand function:*

$$y = y(r, p_c, p_L, I, \ldots). \tag{3.4}$$

Frequently, the term demand function is restricted to the relation between the quantity demanded of any commodity and the price of that commodity only, with other parameters assumed held constant (the Latin phrase *ceteris paribus*— 'other things being equal'—is used to denote this assumption). The reasons for this are that price is thought to be a particularly important influence on demand, price is the major instrument within the control of the firm and the ultimate purpose of the analysis is often to explain how prices are determined. Restricting the demand function to one independent variable also allows a graphical representation. It is long-standing tradition in economics (which engineers find frustrating!) that the independent variable price is always plotted on the vertical axis, and the dependent variable (here, number of calls or connections) on the horizontal axis.

In Fig. 3.3 the demand function $y_1(p)$ for year 1 is downward sloping, indicating that (other things being equal), a lower price will generate a higher quantity demanded. Changes in price correspond to moves *along* the demand curve. If incomes were to rise (over time, for example), the whole demand curve would be expected to shift upward, say, to $y_2(p)$ in year 2.

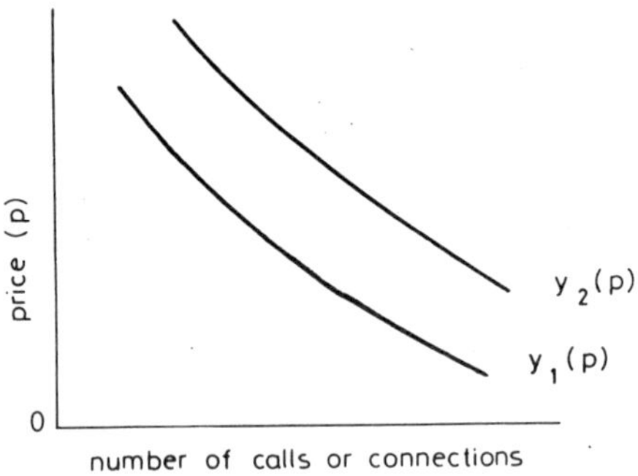

Fig. 3.3   *Demand curves*

3.5   Elasticity of demand

The slope dy/dp represents the rate at which quantity demanded responds to a change in price. A steep curve indicates less response than a flatter one. But the slope itself depends on the units of measurement used, so it is more convenient to express responsiveness in percentage or proportionate terms, independent of units. We define the *elasticity of demand with respect to price (e)* as the proportionate change in quantity demanded which would be generated by a unit proportionate change in price. Formally,

$$e = \frac{dy}{y} \bigg/ \frac{dp}{p} = \frac{dy}{dp}\frac{p}{y} \tag{3.5}$$

Assuming that demand decreases as price increases, this price elasticity of demand will be negative. Thus an elasticity of $e = -2$ means that a 1% increase in price will lead to a 2% decrease in quantity demanded. Demand is said to be *elastic* if $|e| > 1$ and *inelastic* if $|e| < 1$, and to have *unit elasticity* if $|e| = 1$.

It is revealing to express total revenue in terms of price elasticity of demand in order to examine the relation between revenue and price. Total revenue R as a function of price equals price times demand at the price:

$$R(p) = p\, y\,(p) \tag{3.6}$$

The proportionate change in total revenue per unit change in price is thus

$$\frac{dR}{dp}\frac{p}{R} = \left[y + p\frac{dy}{dp}\right]\frac{p}{py} = 1 + e \tag{3.7}$$

If price is increased, revenue will *rise* if demand is inelastic, will be *unchanged* if demand is unit elastic and will *fall* if demand is elastic. In the latter case, revenue could be increased by *lowering* price.

In a survey of 90 US public utility commission decisions over the period from 1937 to 1946, Troxel (1948) found that only two indicated any allowance for buyer responses to price changes. Unless demand is completely inelastic ($e = 0$), this procedure will lead to financial difficulties. A telephone company ordered to reduce prices in the face of falling costs, as was generally the case in the postwar period, was not likely to object to the consequent increase in traffic and revenue. During the last decade, however, price rises to combat inflation and increased usage have been more common, and ignoring elasticity leads to inadequate revenues. This point is nowadays generally appreciated. For example, one US telephone company computed an elasticity of $-0\cdot32$ and reported as follows:

> If (the Company) were to increase the price of intrastate toll messages by 15%, the volume of intrastate toll messages per day per unit would decrease $4\cdot41$%. Toll revenues would not increase by 15% because of consumer reaction to the rate change. The net increase in revenue based on 1973 data would be only $9\cdot93$% which is about 66% of the desired level. To realise a 15% increase in revenue, the price of toll should be increased by about 23%.*

The US state regulatory commissions have essentially accepted the principle of demand elasticity (or 'repression', as they sometimes call it), but so far have been unwilling to accept the specific numerical estimates of the telephone companies (Lowry 1976, Florida Public Service Commission, 1975, pp. 21–2).

Elasticity of demand need not be constant at all points on the demand curve. For instance, with a linear demand curve $y = a - bp$, elasticity is given by

$$e = \frac{-bp}{a-bp} \tag{3.8}$$

---

*These calculations do not seem exactly consistent. An elasticity of $-0\cdot32$ implies that a price increase of 15% would decrease traffic volume by $15 \times 0\cdot32 = 4\cdot8$% and increase toll revenues by $15 \times (1-0\cdot32) = 10\cdot2$%. Thus to realise a 15% increase in revenue a price increase of $15/(1-0\cdot32) = 22$% would be required.

which increases from zero to (minus) infinity as p increases from zero to a/b (at which level demand falls to zero). Only if the demand function is linear in logarithms, i.e.

$$\log y = \log a - \beta \log p \tag{3.9a}$$

or

$$y = ap^{-\beta} \tag{3.9b}$$

is elasticity constant at all points on the curve, since

$$e = \frac{dy}{y} \Big/ \frac{dp}{p} = \frac{d(\log y)}{d(\log p)} = \beta$$

Log-linear curves are frequently used in empirical estimation because they generally give a good statistical fit and it is easy to read off the elasticity as the estimated coefficient.

The concept of elasticity of demand is equally applicable to other explanatory factors. Thus, *elasticity of demand with respect to income* (abbreviated to income elasticity) is defined as the percentage change in quantity demanded generated by a unit percentage change in income (assuming that price is held constant). Income elasticity will usually be positive for telecommunications, as we shall see, although it may be negative for goods (such as travel by bus) which become less attractive as income increases. Such goods are called *inferior*.

Similarly, one may define the *crosselasticity of demand* as the percentage change in quantity demanded per unit change in price of some other commodity. Crosselasticities will be positive if the two commodities are viewed as *substitutes* (e.g. letters and telephone calls) and negative if they are viewed as *complements* (e.g. telephone calls and PBX equipment). The term *own price elastcity of demand* is sometimes used to distinguish it from crossprice elasticity.

The concept of elasticity is also used with respect to supply, and indeed wherever one variable is related to another.

3.6   Estimating demand functions by econometric techniques

For planning purposes, the engineers, managers and administrators in the telephone system would like to know the position and shape of the demand functions for the various telecommunications services. One possibility is simply to ask subscribers how they would plan to respond to specified situations. Such questionnaires have always proved notoriously unreliable. What the telephone administration does have available, in more or less detail, is a record of past behaviour of subscribers. This does not constitute a complete map of the demand function at any time, only a series of single points on the demand function, one at each period of time. However, by making certain assumptions, it may be possible to reconstruct the demand function which held in the past.

Broadly speaking, two methods are available, involving the use of either time-series data or *cross-section* data. Both methods require data on consumption and on other factors which are thought to explain the level of consumption. The time-series approach required these data for a single group of subscribers over a period of time. The cross-section approach requires data for different groups of subscribers at a single point of time. If all the explanatory factors have been incorporated, these observations may in principle be interpreted as observations on a single demand function. Econometric techniques have been developed to find the function which best fits the data. Details of these techniques, and discussion of the comparative merits of the two methods, can be found in any elementary statistics or econometrics text (Johnston, 1963). In practice, time-series data are most often used because they are more easily obtainable.

The demand function thus obtained may be used to predict future demand if one is willing to assume that the form of the function will continue to hold, and if one is able to predict the future levels of the explanatory variables. This assumption is by no means always justified, and prediction of explanatory variables may not be any easier than predicting telephone demand directly. It is generally found, however, that demand functions of the kind described provide a great deal more insight and accuracy than do simple forecasting models based on time alone. The National Board for Prices and Incomes (1968, pp. 8-9), reporting on the practices of the British Post Office, put it this way:

> There are two main advantages to be derived from using econometric models. First they make for a better understanding of the problem under consideration because the assumptions on which the models are based have to be made explicit. Second, they provide a means of testing assumptions and evaluating the sensitivity of forecasts to changes in assumptions. We have some evidence to show that the long term forecasts made on the basis of econometric models, which have been developed, are no less accurate than those made by alternative methods, although these models are in their early stages of development and are subject to certain technical deficiencies. The use of econometric models involves no extra assumptions about the

stability of the structure of demand than are made by trend extrapolation methods and it is possible to specify the chances that the forecasts of an econometric model are in error by more than a certain amount. We therefore think that the effort being put into this work should be substantially increased.

3.7   Empirical studies of demand for connections

Lonnstrom *et al.* (1967, p. 19ff) report that empirical research in Sweden shows the following factors to be useful for long-term (10–20 year) predictions of demand for telephones; national income per capita, percentage of population in industry and consumption of newsprint per capita. For example, over the years 1910 to 1958 there is a high correlation between the logarithms of number of telephones per hundred population y and national income per capita at 1959 prices I, namely

$$\log y = -5 \cdot 87 + 1 \cdot 91 \log I \qquad (R = 0 \cdot 99) \qquad (3.10)$$

Elasticity of demand with respect to income is $1 \cdot 91$. Thus, a 1% increase in national income per capita (other things such as prices remaining unchanged) would be expected to lead to a $1 \cdot 91\%$ increase in number of telephones per capita.

Roberts and Ward (1974) discuss a number of regression models for Australia, with two and three independent variables. For example, they estimated the function

$$\log y_t = -5 \cdot 00 - 0 \cdot 25 \log r_t + 2 \cdot 03 \log c_t \qquad (R^2 = 0 \cdot 97) \quad (3.11)$$

where for each year t,

y_t = new demand installations
r_t = price of installation (six months' advance rental plus connection fee)
c_t = private consumption expenditure

Here, price elasticity is equal to $-0 \cdot 25$. A 1% increase in installation price would be expected to lead to a $0 \cdot 25\%$ decrease in new demand for telephones. If private consumption expenditure may be viewed as a convenient measure of consumer income, the model indicates an income elasticity of about 2. A given percentage change in income would lead to double that percentage change in new demand.

Because they are independent of units of measurement, elasticities provide a convenient way of summarising demand functions and of comparing demand functions between different countries (and, indeed, between different products). Table 3.1 provides a selection of econometric estimates of elasticities of demand for connections with respect to rentals and income. The results suggest that demand for connections is very inelastic with respect to rentals (absolute elasticity about 0·2 for businesses and less than 0·1 for residences). Income elasticities are somewhat higher, but the spread of results is so wide as to prevent any firm conclusion being drawn. It should be emphasised that these estimates are not directly comparable because of differences in estimation techniques and in definitions of variables (although all the studies are based on time-series data). We shall also see later that the pricing policy of the telephone administration will to some extent affect the observed elasticity of demand.

Table 3.1   *Elasticities of demand for telephone connections*

Source	Country	Type of Telephone	Elasticity with respect to	
			Rental	Income
Waverman (1974)	Canada	Residential	− 0·07	0·18
Waverman (1974)	Canada	Business	− 0·19	n.e.
Simpson (1969)	UK	Business	− 0·20	0·65
Lonstrom *et al.* (1967)	Sweden	All	n.e.	1 91
Waverman (1974)	Sweden	All	− 0 06	0·56
Davis *et al.* (1973)	USA	All private line	−0·02	0·08
Roberts and Ward (1974)	Australia	All (new) demand	−0·25	2·03

n.e. = variable not estimated

Perl (no date) has reported the results of some demographic analyses of cross-section data on age and household group. He finds that price elasticity of demand for connections decreases with income and, more surprisingly, with age of the head of the household.

3.8   Empirical estimates of demand for telephone calls

Telephone organisations have employed a variety of techniques to forecast the number of calls. One model which has been found useful by the British Post Office is described by Fox (1973). Based on annual data from 1958/59 to 1971/72, the equation is

$$\nabla \text{ trunk calls} = -0\cdot18 \ \nabla \text{ average price} + 1\cdot65 \ \nabla \text{ TFE} \qquad (3.12)$$
$$+0\cdot62 \ \Delta\% \text{ STD} \qquad + 0\cdot41 \text{ residual connections}$$
$$(R^2 = 0\cdot81)$$

Here ∇ denotes percentage difference between successive years, while Δ denotes the simple difference in percentages. The result shows that a 1% increase in average price of a trunk call leads to a $0\cdot18\%$ decrease in the number of trunk calls made, i.e. price elasticity $= -0\cdot18$. Elasticity of demand with respect to total final expenditure (TFE), another measure of national income, is $1\cdot65$ and elasticity with respect to the number of residential connections is $0\cdot41$. Finally, an increase of 1% in the percentage of self-dialled (STD) calls leads to a $0\cdot62\%$ increase in the number of trunk calls.

The US telephone company referred to in Section 3.5 developed a toll demand model which related total toll messages per telephone in each month to that same variable in the previous month, per capita personal income in the state, an index of the price of toll calls and an index of business activity in the state (namely, the consumption of electric power). The purpose of the last independent variable is presumably to identify a determinant of business demand comparable to the income of personal customers and to prevent this factor influencing the estimate of price elasticity; whether consumption of electric power is easier to forecast than telephone usage is not clear. The model, tested over the period 1966–1973, showed a 'habit elasticity' of $0\cdot33$, an income elasticity of $0\cdot64$, a price elasticity of $-0\cdot32$ and a business activity elasticity of $0\cdot24$. Lowry (1976) gives several further examples of work by New York, New England, and Pacific Northwest Bell Companies.

Mention should also be made of the Bell System's Long Distance Interstate (LDI) demand model, developed by the Long Lines Company in 1967 and currently used to estimate revenue and traffic effects of proposed rate changes. This model involves a combination of econometric and simulation techniques, as described in a supplement to the testimony of Betteridge (1973).

A selection of estimates of price and income elasticities are set out in Table 3.2. All these studies are based on time-series data. Once again, it should be borne in mind that the data and definitions vary, and often the estimating equations contain other variables not reported here. To the extent that a pattern emerges, it seems to be as follows:

Table 3.2   *Demand for telephone calls*

Author	Country	Dependent variable	Elasticity with respect to	
			Price	Income
Turner (unpublished	UK	Local calls	− 0·06	0·20
Simpson (unpublished)	UK	Trunk calls (full rate)	− 0·52	1·76
Waverman (1974)	UK	Inland trunk calls per telephone	− 0·63	1·11
Waverman (1974)	Canada	Residential revenue per telephone	− 1·20	0·99
Waverman (1974)	Canada	Business revenue per telephone	− 1·20	0.45
Waverman (1974)	Sweden	Trunks calls per telephone	− 0·29	n.s.
Waverman (1974)	Sweden	Trunk calls per telephone	n.s.	1·02
Davis *et al.* (1973)	USA	Local	− 0·21	0·25
		Toll	− 0·88	0·83
		Private line	− 0·74	n.e.
		WATS	− 0·14	n.e.
Lago (1970)	International	International calls	− 1·25	n.e.
Naoe (1974)	Japan	International calls	− 2·28	n.e.
Drew (1973)	UK	International messages (calls and letters)	− 0˙86	n.e.

n.e. = not estimated

n.s. = not statistically significant

(*a*)   Price elasticity of demand is very low for local calls (less than −0·2), somewhat higher, but still inelastic, for trunk calls (−0·3 Sweden, −0·5 to −0·6 Great Britain, −0·9 USA) and elastic for international calls (−1·25 to −2·3). Elasticity thus seems to increase with distance and price, as one might expect.

(*b*)   The results for Canada show about equal price elasticities of demand by residential and business subscribers. This is surprising, since business demand is usually thought to be significantly less elastic. It is also surprising that these estimates show overall demand to be elastic (−1·2), whereas the other results just mentioned show inelastic demand.

(*c*)   Income elasticity of demand is very low for local calls (just over 0·2), but is usually elastic for trunk calls (varying from 0·9 to 1·8).

(*d*)   Income elasticities for residential subscribers in Canada are about twice as high as for business subscribers.

See, however, Table 3.3 and comments in Section 3.10, and recent work by Auray (1978) and Doherty (1978).

One would expect that a higher price would lead to calls of shorter duration. Gale (1976) has estimated an average elasticity of duration with respect to price of $-0\cdot14$ for intrastate toll calls (up to 124 miles) based on cross-section data. This price elasticity is higher for business than for residence subscribers, and higher for collect (reversed charges) and personal calls than for station calls. It should be noted that price is calculated as 'average price' per minute rather than 'marginal price' of an additional minute; strictly speaking, it is the latter which is relevant for the decision about duration of call. Elasticity of duration with respect to income is estimated at $0\cdot32$.

A great deal more work has been done by statisticians in telephone organisations than is available in the open literature. For example, there have been some studies in Britain of the effect of postal rates on volume of telephone calls, and P. and B. Brandon of Bell Telephone Laboratories have worked on demographic analyses of calling patterns.

Some of the models have now been around long enough for one to assess their accuracy in forecasting. For example, Simpson's first model on full-rate trunk calls in Great Britain, published in 1965, forecast a growth in trunk traffic of 245% from 1963 to 1974. As it turned out, the forecast was accurate. The form of the model used by the British Post Office today is not very different from that first model. By contrast, Simpson reports* that none of his residential connection models has lasted more than a year! It would be interesting to know how robust other models have turned out to be.

3.9   Perceptions, commitments and lagged response

Although the demand function purports to measure the response of consumers to changes in various parameters, it should not be assumed that this response necessarily takes place immediately. There are basically two reasons why the response might be delayed, and hence the full effects of a change not felt for some time.

First, customers can respond only to prices and opportunities which they perceive. If they are not aware that offpeak prices have been reduced, it is not surprising if they do not shift their calls to the evening. On 1 July 1970 morning peak rates were introduced into Britain for the first time, at a level 25% higher than standard (afternoon) rates. There

*Private communication

was very little advertising of the fact, it did not show up on the subscriber's bill, and there was no immediately observable impact on traffic patterns. As subscribers have gradually come to learn of this price differential, and as telephone rates have risen generally, making it more worth while for the subscriber to familiarise himself with the rates, there has subsequently been a significant shift of traffic from morning to afternoon.

In similar vein, W. Simpson* has put forward the hypothesis that price elasticity of trunk calls depends on the billing mechanism. With Subscriber Trunk Dialling (STD), customers in Britain no longer receive itemised bills; hence they are not informed of the prices of individual calls. Firms (and fathers!) cannot easily monitor the use of the telephone. This may be why the price elasticity of UK trunk calls has been falling over time, why a lower elasticity is observed in the UK than in the USA and why the STD coefficient in the demand function of eqn. 3.12 appears to be so high.

One of the functions of advertising is to draw the attention of customers to new commodities, new prices or new aspects of existing commodities. In the early 1970s the British Post Office invented the following 'jingle' to popularise offpeak rates in their television advertisements:

> It's so cheap to call your friends
> After six and at weekends.

A bird called Buzby has since been pressed into service to deliver similar messages.

A person who has never used a telephone is unlikely to rent one until he has gained experience at the office or at the homes of neighbours, and has appraised the benefits it could bring. This is particularly important in developing countries, where a single telephone in a village may be a useful investment (by the telephone administration) for purposes of developing 'the calling habit', even if the immediate revenue does not cover expenses.

Boarts (1964) makes the interesting suggestion that relatively low levels of telephone development in certain New England metropolitan areas in 1958 could be explained by the comparatively short period of time for which those areas had experienced high income levels. Telephone development had not yet caught up with income growth. Assuming that there was no shortage of capacity in the telephone system (in which case the level of development would reflect conditions

---

*Private communication

of supply rather than demand), this may be another instance of the phenomenon we have been discussing. Potential customers had not yet realised the merits of becoming subscribers. On the other hand, it may equally be that they did appreciate the benefits of a telephone, but had higher priorities for expenditure at their new levels of income (for example, on other consumer durables).

A second explanation for time lags in response is that subscribers develop personal habits or business commitments based on an initial pattern of prices or incomes. A rapid response is not necessarily desirable or profitable; in extreme cases there may be no response at all. For example, a subscriber might have found it worth while to join and pay the connection fee at the original (low) rental; if he had the decision again at the new (higher) rental, he would not find it worth joining, but since the connection fee has been paid, he finds it worth while to continue subscribing. The changed rental affects only new subscribers. Again, the gradual reduction in costs of long-distance data transmission has stimulated the growth of the remote-access computer business, but it takes time to establish such a business, especially since many potential customers have already purchased their own computer; the full effect will not be felt until these computers become obsolescent.

Table 3.3   *Alternative calculations of residential long-distance elasticities for Canada*

	Income		Price	
	Short run	Long run	Short run	Long run
Dobell *et al.*	0·20	1·30	− 0·30	− 1·90
Kotowitz and Waverman	1·25	1·25	− 0·91	− 0·91

[Source: Kotowitz and Waverman (1973), Table 1]

One might, therefore, expect a lower immediate response than after some time has elapsed for adjustment. Economists use the terms *short-run* and *long-run* to distinguish between such situations. Dobell *et al.* (1972) made ten sets of estimates of short-run and long-run price and income elasticities for Canada. They found that the demand for telephone services was characterised by strong habit formation which was reflected in a substantial difference between short-run and long-run elasticities (long-run elasticities being greater, as one would expect). However, Kotowitz and Waverman (1973) have argued that the demand functions in that study were misspecified. The results implied implausibly long adjustment times (up to 20 years). Their own studies suggested that residential subscribers adjust their demand for long-distance calls within one year of a price or income change, and that short-run and long-run elasticities are not significantly different. The two sets of elasticities are shown in Table 3.3.

3.10   Crosselasticities of demand by time of day

The number of telephone calls made does not usually show a marked variation from week to week or month to month (except in holiday resorts). Demand is generally lower at weekends and on holidays because of the reduction in business calls. The most significant variation is *within* the day. Fig. 3.4 shows a typical daily load curve in Great Britain in October, 1967 (before the morning peak rate was introduced). The peaks (or busy-hours) occur in midmorning and midafternoon. Observe the spare capacity at other times; the weekday 24 h load factor is only about 25%. Many developing countries attain a much higher degree of capacity utilisation, but only because there is severe congestion which limits usage during much of the day.

We might be interested in knowing how easy it is to induce daytime customers to shift their calls to night time by means of lower prices then. This can certainly be done; in fact, in the USA in the late 1960s the evening prices were reduced to such an extent that the daily peak (the busy-hour) on heavy-traffic long-distance routes (e.g. between Chicago and New York) was inadvertently shifted from the morning to early evening.

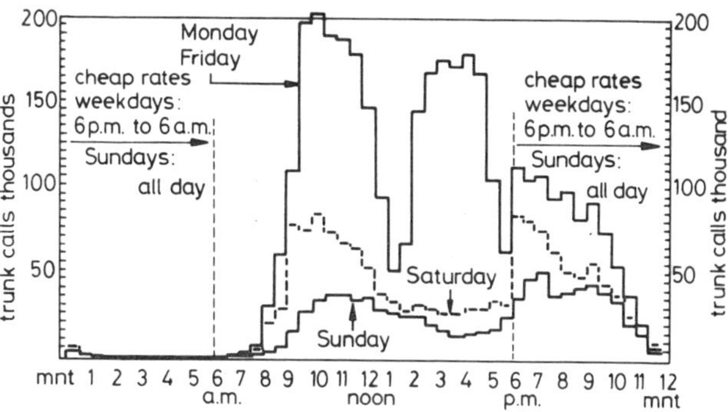

Fig. 3.4   *Half-hourly incidence of trunk calls, October 1967*
Distribution based on operator controlled calls [Source: National Board for Prices and Incomes (1967), Appendix 9]

As noted earlier, the elasticity of demand for telephone calls at one time of day with respect to price at another time of day is called the *crosselasticity of demand*. Little empirical work has been published on crosselasticities. One reason, no doubt, is that price differentials between peak and offpeak hours have been rather limited, and have changed only infrequently, so that few data are available for testing. In view of this it seems worth while to present some calculations based on data given by Auray (1969, exhibit 5) concerning the effects on US interstate traffic of an after-9 p.m. rate reduction in April 1963. Aggregating person and station calls, we may present his data as in Table 3.4. For traffic in the 507–925 milage band, night-time (reduced rates) traffic has an elasticity of $-0\cdot55$ with respect to its own price, whereas daytime traffic (full rate) has a crosselasticity of $0\cdot12$ with respect to the night price. For traffic in the longer-distance band, these elasticities are respectively twice and three times as high. These data suggest there is considerable scope for pricing policies designed to shift traffic between different times of day — at least, if constraints on revenue are not too severe. Hopley (1978) and Auray (1978) have since published the results of more recent research on elasticities of local and intrastate calls, respectively.

Table 3.4 *Crosselasticity of demand by time of day*

Time of Call	507 – 925 mile band			926 – 3006 mile band		
	Change in number of messages	% of total	Own- or crosselasticity with respect to reduced rate	Change in number of messages	% of total	Own- or crosselasticity with respect to reduced rate
Full rate	$-0\cdot28$ M	$-2\cdot0$	$0\cdot12$	$-0\cdot48$ M	-7	$0\cdot37$
Reduced rate	$+0\cdot46$ M	$+9\cdot3$	$-0\cdot55$	$+1\cdot08$ M	$+19\cdot1$	$-1\cdot01$
Total	$+0\ 18$ M	$+0\cdot95$	$-0\cdot56$	$+0\cdot60$ M	$-4\cdot79$	$-0\cdot25$
Reduced-rate price change		$-17\cdot0$			$-19\cdot0$	

[Source: Auray (1969), Exhibit 5, modified by the present author]

3.11 A reservation

It is tempting to think of successive econometric studies as gradually providing more and better information to pin down the exact levels of demand elasticities. This is, after all, the procedure of biologists and physicists working to establish the properties of organisms and metals. Yet it may be argued (Hayek, 1952, Shackle, 1972), that the fundamental difference between the social and natural sciences is precisely this: that in the former there are no permanent empirical constants such as there are in the latter. Economics and sociology cannot hope to discover quantitative laws of behaviour as biology and physics can. To be sure, empirical information about the elasticity of demand at some particular periods in the past may be extremely helpful in predicting and making plans for the future. Nevertheless, the constantly changing environment and the ability of men to create new products and opportunities mean that the extrapolation of past behaviour must always be treated with caution.

Chapter 4

Production functions

4.1 Outputs and inputs

The process of production involves the carrying out of certain *activities* in order to produce certain *outputs*. These activities require certain *inputs*, or *factors of production*. A *production function* is a schedule relating outputs to inputs: it specifies the maximum output (or outputs) which can be produced by any given set of inputs. A *cost function* is a schedule relating outputs to costs of inputs: that is, it specifies the minimum total cost of producing any given output (or outputs). It has been suggested (Ellis, 1975) that economists prefer working with production functions and engineers prefer to use cost functions—for example, when analysing the meaning and extent of 'economies of scale'. Although the two concepts are evidently closely related, they can provide different insights and it will be worth studying them separately.

In telecommunications, the ultimate outputs are telephone calls, data transmission or other forms of communication, plus a variety of additional subscriber services. The way in which the inputs are categorised depends upon the purpose of the analysis. For operating and planning purposes, telephone administrations may find it useful to think of exchange operators, repairmen, accountants, toll offices, microwave systems, electricity etc. as different kinds of inputs. Alternatively, a government concerned about levels of employment and investment in the economy as a whole might think of the inputs simply as labour, capital, land etc.

Economists have been regrettably lax in specifying how knowledge of such production functions is obtained, and whether they embody facts about the past or conjectures about the future. In most conventional expositions, no distinction is drawn between these two possible interpretations, since knowledge of the production function is some-how 'given' to all producers, and unexpected outcomes or technological change are ruled out by assumption. We shall try to be more precise.

For purposes of future planning, the production function is a hypo-thetical or conjectured relationship: it reflects what technical opportu-nities the decision maker believes are available to him. If there is some method of production which he does not know, obviously that method will not enter the production function which forms the basis of his decision. It is also possible that the decision maker has made incorrect forecasts about how well a particular machine will work, or whether production will run smoothly according to plan. If so these errors are not revealed at the planning stage and are, therefore, embodied in the conjectured production function.

Of course, in forming this production function, the decision maker will be guided by his experience of the past, and in particular by the previous levels of output attained from various combinations of inputs. At the end of each period (*ex post*) he will be able to compare what actually happened with what he expected at the beginning of that period (*ex ante*). The production functions estimated by econometric techniques, as described later in this chapter, are attempted recon-structions of what, in hindsight, the production function was 'really' like, as far as one can judge from the point(s) actually observed.

The production function for a single output with many inputs may be formally represented as

$$Q = f(x_1, x_2, \ldots, x_n) \tag{4.1}$$

where Q denotes the maximum amount of output which may be produced in a given period of time by using amounts $x_1, x_2, \ldots, x_n$ of inputs $1, 2, \ldots, n$.

Such a function can be illustrated graphically by assuming only two factors of production, conventionally called capital K and labour L, so that $Q = f(K, L)$. It is assumed that a given level of output can be pro-duced in different ways by using different combinations of the two inputs. Fig. 4.1 shows the various combinations of capital and labour which might be used to provide two different levels of output—say, 5000 and 10 000 erlangs busy-hour capacity. The *capital/labour* ratio at any point is measured by (the inverse of) the slope of a line from the origin to that point.

The *labour-intensive* method (i.e. a method with a high ratio of labour to capital) might involve a manual or Strowger exchange, a *capital-intensive* method might involve an electronic switching system. A crossbar system would rank in between. In practice there might be only a limited set of alternatives, but for ease of exposition the lines are drawn as continuous. Each line is called an *isoquant* or 'equal product curve' (in other disciplines, such curves are sometimes called 'level curves').

Notice that the isoquants are drawn downward sloping and convex to the origin. If less labour is used, more capital must be substituted for it. However, the rate at which capital needs to be substituted increases as more substitution takes place: substitution thus becomes more and more difficult. The slope of the curve measures the *marginal rate of substitution* between the two factors, and the convexity property is referred to (somewhat strangely) as a diminishing marginal rate of substitution.

4.2   Returns to scale

A production function is said to exhibit *constant, increasing* or *decreasing returns to scale* according to whether a given increase in *all* inputs gives an equal, greater or smaller increase in output—that is, according to whether output is unit elastic, elastic or inelastic with respect to inputs. Formally, suppose that multiplying all inputs by a positive constant λ leads to an increase in output of λ^a, where a is called the *scale coefficient*. In terms of the production function of eqn. 4.1, we have

$$f(\lambda x_1, \lambda x_2, \ldots, \lambda x_n) = \lambda^a f(x_1, x_2, \ldots, x_n) = \lambda^a Q \qquad (4.2)$$

This production has constant, increasing or decreasing returns to scale according as $a = 1$, $a > 1$ or $a < 1$. The terms *economies of scale* and *diseconomies of scale* are synonymous with increasing and decreasing returns to scale, respectively.

As Figure 4.1 is drawn, to double output (i.e. busy-hour capacity) from 5000 to 10000 erlangs, it is necessary to double the amount of labour and capital used in the crossbar exchange, whereas the Strowger exchange requires more than double the inputs, and the electronic system requires less than double the inputs. Thus the production function shown here displays constant returns to scale in the region corresponding to the crossbar system, increasing returns to scale for the electronic system and decreasing returns to scale for the Strowger system. This assumption is for purposes of exposition only, for it is arguable that even for Strowger equipment a large exchange produces higher (or as high) output per bundle of inputs as a small exchange does. The relationship can be alternatively exhibited as in Fig. 4.2,

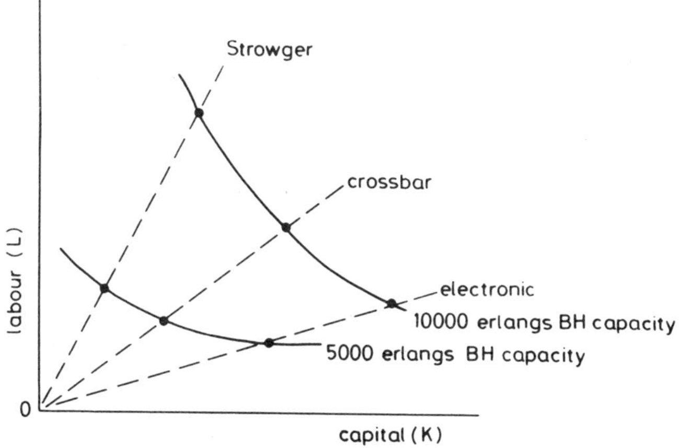

Fig. 4.1   *Isoquants of the production function*

where the horizontal axis refers to some index of all inputs assumed held in the same proportions to each other.

On reflection, it may seem surprising that constant returns do not prevail everywhere. If all inputs are doubled, why is not output precisely doubled? In theory, this must be the case. In practice, it is extremely difficult to discover situations where all inputs can be increased in the same proportion:

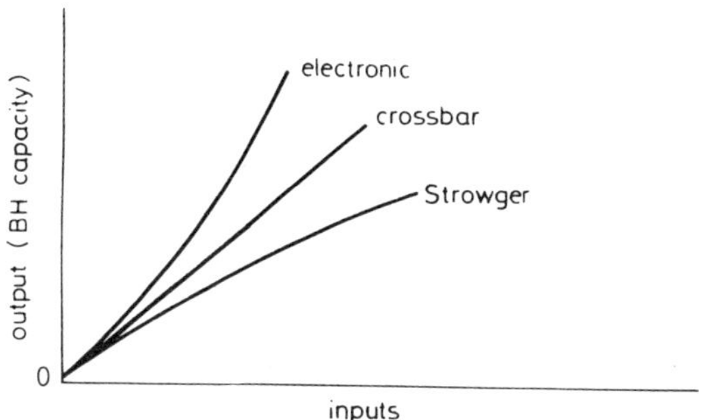

Fig. 4.2   *The production function and returns to scale*

(i)   It may not be possible to obtain further supplies of each factor at exactly the same quality. For example, if the telephone company wished to double the number of operators, it would no doubt have to accept many who were less capable than its original employees.

(ii)   Some inputs are available only in indivisible 'lumps'. One cannot install fractions of a circuit, nor employ half a manager all the time (which is not the same as a whole manager half the time!)

(iii)   There are certain technical relationships, such that area increases as the square of radius, which automatically prevent constant returns. In telecommunications, the 'law of large numbers' means that doubling the number of circuits more than doubles the traffic capacity at a specified grade of service.

(iv)   It is impossible to specify all the factors which are required to produce each output, and some factors which are, in fact, relevant may have been overlooked and not increased.

Thus, in practice, changes in output are invariably accompanied by changes in the proportion in which inputs are combined. It is not possible to determine by theory alone whether a production function will display increasing or decreasing returns to scale with respect to those inputs which are actually measured and taken into account.

4.3   Diminishing returns or variable proportions

We have already observed that, at a given level of output, it becomes more and more difficult to substitute one factor for another. We now wish to examine how output is related to varying amounts of one input when other inputs are held constant. This relationship is known as the *returns to a factor or production*. Suppose we take a manual telephone exchange of given size and equipment (i.e. we fix the capital K) and examine how the capacity to handle traffic (at specified grade of service) varies with the number of operators (i.e. we vary the labour L).

Fig. 4.3 shows the likely relationship. The first few operators each provide a significant increase in output, but subsequent operators are surplus to the capacity of the exchange, and indeed, too many operators can even reduce output through overcrowding.

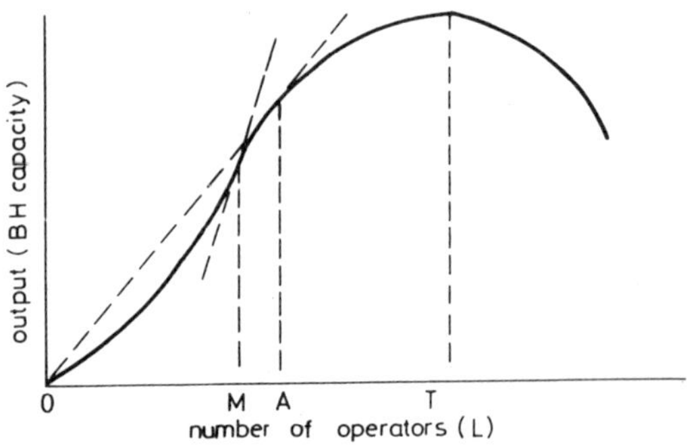

Fig. 4.3   *Variable proportions in a manual exchange of fixed size*

Maximum total output in this exchange (of fixed size) is obtained with T operators. Average output per operator, also called average return to operators, or average product per operator, is given geometrically by the slope of a line from the origin to the point on the curve corresponding to a particular number of operators. Here it attains its maximum with A operators. Marginal output, also called marginal return to operators, or marginal product, is the change in total output generated by an additional operator, given by the slope of the curve at that point, and reaches its maximum even earlier, at M operators.

We have thus four distinct ranges of the labour input (operators). In the range OM, both average and marginal returns to labour are increasing; in the range MA average return is still increasing, but marginal return is decreasing (and at OA marginal and average returns are equal); in the range AT both average and marginal returns are declining, but still positive, so total returns are increasing; finally, in the range beyond T, marginal return is negative and both average and total returns are decreasing.

Obviously, it would be inefficient to employ more than T operators, when fewer operators could actually produce more output. What is not obvious from the Figure, but is nonetheless true, is that it will never be efficient to produce in the range OA either, because output could be increased by reducing one of the factors currently held fixed (e.g. by building a smaller exchange). The only range in which efficient production takes place is AT, a range in which the marginal return to each factor is declining. The optimal position within this range is determined by relative costs of each factor, which we deal with in the next chapter.

4.4 Empirical studies of returns to scale in telecommunications

On intuitive grounds, what can be conjectured about the existence of economies or diseconomies of scale in the various different elements of a telecommunications system?

It is sometimes said that local exchanges encounter diseconomies of scale because the number of possible connections between subscribers increases with the square of the number of subscribers. However, the fact that not all subscribers wish to use their telephone at the same times makes possible certain economies, as indicated in remark (iii) of Section 4.2. For example, at $0\cdot5\%$ grade of service, 1 trunk provides a capacity of $0\cdot005$ erlangs, 10 trunks provide 4 erlangs and 100 trunks provide 81 erlangs (Morgan, 1976, Appendix H). This topic is dealt with further in chapter 12.

It seems likely that exchange switching equipment offers a small degree of increasing returns to scale. To double the output of a Strowger exchange it is necessary to double the switching equipment, but not the power plant, building or land. With crossbar, and particularly electronic systems, the common-control equipment does not need to be doubled to double output.

Economies of scale are probably most significant in transmission, where there are enormous indivisibilities. Much terminal equipment has to be increased with output, but the same radio relay towers and coaxial cables may be used. It follows that economies of scale are greatest on the longer-haul routes, where the terminal equipment is a smaller proportion of the total factor inputs.

These conjectures are consistent with the calculations of Mantell (1975*a*), who reports that the US Bell system as a whole has a scale coefficient of 1·16 ± 0·13, but that, taken separately, the Long Lines company and the Operating Companies exhibit coefficients of 1·26 and 1·10, respectively. It should be noted, however, that Vinod (1975*a*) has expressed reservations about some of the statistical procedures used, and that McElroy (1975), using several techniques, failed to find any clear pattern of economies or diseconomies of scale.

It may be useful to examine in a little more detail two empirical studies of production functions for telecommunications systems taken as a whole. These studies are based on time-series data for a single organisation. This method presents two problems. If all the variables are growing over time at about the same rate (referred to statistically as 'multicollinearity'), it is difficult to distinguish the effects of each variable separately. The estimated coefficients have a large variance. Second, it is necessary to distinguish between economies resulting from increased size of output and economies resulting from technological progress. To do this, a separate variable representing technical progress is usually introduced; the difficulty is to find a variable which will measure this accurately.

Dobell *et al.* (1972) found that in Bell Canada the capital/labour ratio (measured in dollars of net capital stock per man-hour) increased from 12·58 in 1952 to 41·75 in 1967. The capital/output ratio increased from 2·33 in 1952 to 3·75 in 1961 then decreased to 3·46 in 1967 (these ratios may be interpreted visually in diagrams such as Figs. 4.1 and 4.3, respectively). According to the authors, the most plausible explanation for these trends was the heavy capital investment in direct-dial equipment in the late 1950s, which replaced operator-switching systems by machine-switching systems.

Using econometric (time-series) techniques, the authors estimated the following production function:

$$\log V = -0\cdot25 + 0\cdot705 \log L + 0\cdot405 \log K + 0\cdot010\, D \qquad (4.3)$$

where

    V = value added (revenue less expenses) per year
    L = man-hours employed per year
    K = net capital stock
    D = percentage of station calls self dialled

The coefficient $0\cdot705$ may be interpreted as the elasticity of output with respect to labour input. It says that a 1% increase in labour input would yield a $0\cdot7\%$ increase in the value of output. Similarly, a 1% increase in capital employed would yield a $0\cdot4\%$ increase in the value of output. These elasticities are essentially the marginal returns to (or marginal productivities of) labour and capital, expressed in a form which is independent of units. The variable D is used as a proxy for technological progress; self dialling permits the handling of a larger number of calls without corresponding increases in telephone operators. The coefficient of $0\cdot01$ indicates an annual rate of technological progress equal to the rate at which the percentage of self-dialled calls is increasing, which over the period of study was about $4\cdot75\%$ per year.

A similar study was carried out for the USA by Davis *et al.* (1973), who estimated for the period 1954–1970 the production function

$$\log V = 2\cdot504 + 0\cdot549 \log L + 0\cdot545 \log K + 0\cdot053 \text{ time}$$
$$(R^2 = 0\cdot999) \qquad (4.4)$$

This equation suggests that the marginal productivity of labour in the USA is lower than in Canada, and the marginal productivity of capital slightly higher. The passage of time was taken as a proxy for technical progress and the coefficient of $0\cdot053$ suggests that, even if the value of the inputs remained unchanged, output would increase at the rate of $5\cdot3\%$ per annum. This may be explained either by *'learning by doing'*, in which experience increases the efficiency with which a given set of resources is managed, or by technical improvements in the quality of the capital equipment.

These aggregate production functions are easy to check for economies of scale. Log–linear functions exhibit increasing or decreasing returns to scale according to whether the sum of the coefficients is greater or less than unity. For example, if $f(K,L) = AK^{\alpha}L^{\beta}$, and λ is some positive constant,

$$f(\lambda K, \lambda L) = A (\lambda K)^a (\lambda L)^\beta = \lambda^{a+\beta} f(K, L) \qquad (4.5)$$

The Canadian data exhibit mildly increasing returns to scale, for if both labour and capital inputs are increased by a 1%, the total value of output would increase by $0 \cdot 705\% + 0 \cdot 405\% = 1 \cdot 119\%$. Further calculations led the authors to suggest that these economies of scale are not to be found in the total wage bill, even though average hours of labour input for the most part fall with increasing output. Rather, the economies are to be found in falling raw materials requirements per unit of output. The US data indicate a similar degree of increasing returns ($0 \cdot 549\% + 0 \cdot 545\% = 1 \cdot 09\%$), although elsewhere Mantell (1975*a, b*) and Vinod (1975*b*) obtain somewhat higher estimates of $1 \cdot 16\%$ and $1 \cdot 20\%$ for the same country.

Such studies must be interpreted with care. The estimated equations measure *value* of output, not physical output. This is necessary in order to add up different kinds of output (i.e. calls, main stations, PBXs etc.). It also means that value of output can be changed simply by a change in prices of outputs. Similarly, net capital stock is a variable whose magnitude depends on depreciation policy as well as on physical capacity. Indeed, the capital variable would decrease, even if there were no change in quality or quantity of physical equipment, simply as a result of the book depreciation embodied in the firm's accounting policy.

It should also be emphasised again that, as with demand functions estimated by time-series methods, one does not have a set of points observed on the same production function, but rather one point from each of a set of functions, one for each year. If these functions could be guaranteed not to shift or change over time, there would be no difficulty. In practice, one attempts to identify the likely changes, and eliminate or explain their effect, but one can never be satisfied that this has been completely achieved. It may also be that decisions about input combinations which seemed efficient at the time subsequently turned out to be mistaken. Production may appear to take place in inefficient regions. The observed output may lie not on but below the 'true' production function.

Chapter 5

Cost functions

5.1 The nature of cost

The concept of cost appears to present no difficulty to the layman: the cost of anything is simply the money that has to be paid for it. Economists have developed a more general notion: the cost of anything is the value of what has to be sacrificed to obtain it. If money has to be paid out, what that money could have bought elsewhere has been sacrificed. There may still be a cost even if no money is paid. To go for a walk may mean sacrificing an afternoon spent watching a football game; using a satellite to transmit television pictures precludes the use of those channels for telephone conversations. This is the notion of *opportunity* (or *alternative*) *cost* (Alchian, 1968, Alchian and Allen, 1974).

The nature and implications of opportunity cost are by no means a settled matter among economists. Indeed, differences of opinion on this matter are at the heart of the rethinking of welfare economics. We discuss this matter further in chapter 14. For the present, we shall attempt to make the exposition as straightforward as possible by means of several simplifying assumptions. Their purpose is to enable us to use money as a convenient first approximation to cost. We assume the following:

(i)   Money outlay is the only relevant consideration—there are no 'non-pecuniary' considerations such as prestige, protection of employment, fairness to suppliers etc.

(ii)   Beliefs about the money outlays needed for decisions are held with certainty—there is no need to weigh up a pessimistic view against an optimistic one.

(iii)   All money payments are made at a single point in time—or at least the period of time over which money is paid out is sufficiently short that no account needs to be taken of the relative value of money at the beginning of the period compared with the value at the end.
(iv)   There is no shortage of cash—no other projects are competing for a limited budget.
(v)   There are no existing stocks, capacity, work force or other assets—all inputs have to be purchased anew.

With these assumptions we may define cost as equivalent to money outlay. Assumptions (iv) and (v) will be relaxed during the course of this chapter. The value of money over time—assumption (iii)—is dealt with in chapter 7. Assumptions (i) and (ii) are discussed in chapter 14.

5.2   Cost functions and the method of production

A cost function is a schedule specifying the minimum cost of producing any given output(s). It is intimately related to the production function, for the cost of any given output is calculated by choosing the cheapest bundle of inputs capable of producing that output (Ferguson, 1971, Ferguson and Gould, 1975). Formally, if there are n inputs with constant unit prices $p_1, p_2, \ldots, p_n$, the cost $C(Q)$ of producing Q units of output is defined by

$$C(Q) = \min\ (p_1 x_1 + p_2 x_2 + \ldots + p_n x_n) \qquad (5.1)$$

$$\text{subject to } f\ (x_1, x_2, \ldots, x_n) \geq Q$$

Obviously, cost depends on the prices of inputs as well as the quantity of output. It should also be emphasised that a cost function used for planning purposes is a conjecture about the future, not a fact about the past, and it is based on a production function having the same character. We are dealing with *prospective costs* and not with *historical costs* represented by payments made at some time in the past. Of course, as with production functions, the estimates of prospective costs may well be based in part on previous experience, and econometric estimates are precisely an interpretation of historical cost data.
Consider the production function represented by the isoquant map of Fig. 4.1. If the telephone organisation wishes to provide capacity to handle busy-hour traffic of 5000 E, what type of exchange should it install? The answer will depend on the relative costs of capital and

labour. Suppose these inputs can be acquired at prices which do not depend on the quantity purchased. We may then draw a straight budget line, also called an *isocost line*, representing the different combinations of capital and labour which can be purchased for a given sum of money. The slope of this line equals the ratio of the prices of the two inputs. In fact, a set of isocost lines may be drawn, parallel to each other, one for each specified total outlay. Isocost lines nearer the origin correspond to lower levels of cost than those further away. The lowest isocost line which intersects a given isoquant is the line which is tangent to it. The cheapest method of producing any specified output is therefore given by the combination of inputs represented by the point of tangency between the appropriate isoquant and an isocost line. This situation is shown graphically in Fig. 5.1. At the assumed relative prices, it is cheapest to use a crossbar system for an exchange of 5000 E busy-hour capacity, but for an exchange with twice that capacity an electronic system would be cheaper.

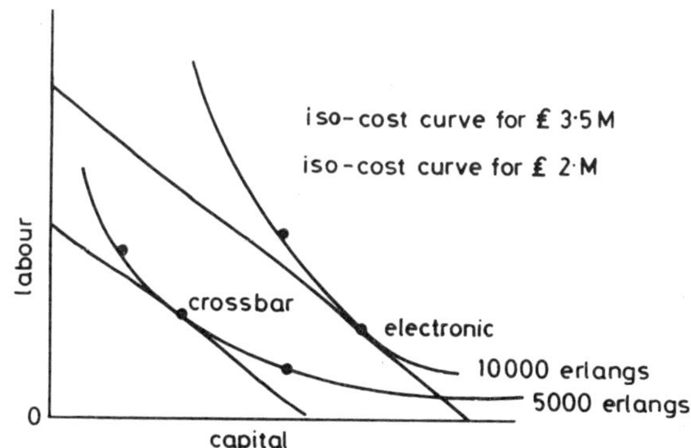

Fig. 5.1   *Least-cost combinations of inputs*

Because the isoquant and the isocost line are tangential at the optimum input mix, it follows that the marginal rate of substitution of capital for labour (given by the slope of the isoquant) is equal to the ratio of the price of capital to the price of labour (given by the slope of the isocost line). At any other point on the isoquant, it would work out cheaper to substitute one input for another, and thereby move along the isoquant until the point of tangency was reached.

It is now apparent how different countries can quite sensibly choose different technologies. In developing countries labour is relatively cheap compared with capital. With the axes labelled as in Fig. 5.1, the isocost lines will be more nearly vertical and the optimal equipment will be labour-intensive, such as Strowger. In countries where labour is expensive, as in the USA, the isocost line will be flatter, so that an electronic system will be cheaper, For similar reasons, one would expect the rise in income over time in any country to lead to a gradual replacement of older labour-intensive (manual) systems by newer capital-intensive (electronic) systems.

5.3   Average and marginal costs

Repeating the calculation of eqn. 5.1 for all possible levels of output Q yields the schedule $C(Q)$ known as the *total-cost function. Average cost* is defined as total cost divided by the volume of output. *Marginal cost* is defined as the rate at which total cost changes with output. Both these costs vary with output. Formally, the *average-cost function*, which we may call $AC(Q)$, is defined by

$$AC(Q) = \frac{C(Q)}{Q} \tag{5.2}$$

If the total cost function is continuously differentiable, the *marginal-cost function*, which we may call $MC(Q)$, is defined as its derivative

$$MC(Q) = \frac{dC(Q)}{dQ} \tag{5.3a}$$

Otherwise, it may be defined as the cost of one additional unit of ouput:

$$MC(Q) = C(Q + 1) - C(Q) \tag{5.3b}$$

What shape do these curves have? A typical example is shown in Figs. 5.2 and 5.3. It is usually assumed that, at first, total cost rises steeply with output as important indivisible inputs are purchased; later, it increases more slowly, since these indivisible inputs do not need to be increased; eventually, it increases steeply again as it becomes more and more difficult to find and combine suitable inputs. The corresponding average- and marginal-cost curves both fall at first, then rise. The marginal curve can be thought of as pulling the average curve up or down towards itself, much as a batsman's latest cricket or baseball score pulls his season's average up or down, depending on whether it is above or below that average. Consequently, if the average-cost curve does rise, the marginal-cost curve always cuts it from below at its minimum point. Equivalently, where average cost, given by the slope of a line from the origin to the total-cost curve, reaches its minimum point it is equal to

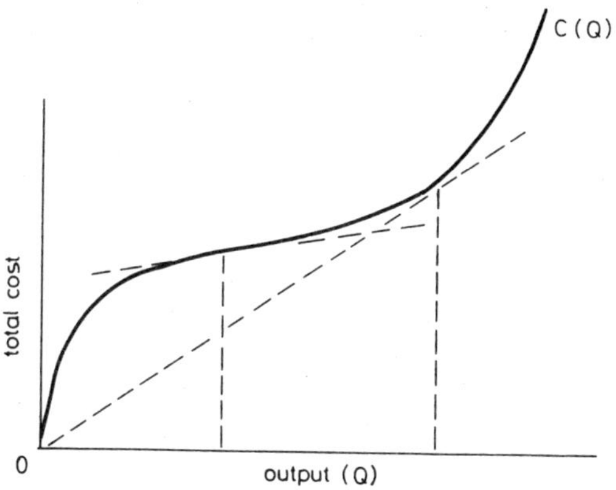

Fig. 5.2   *A total-cost function*

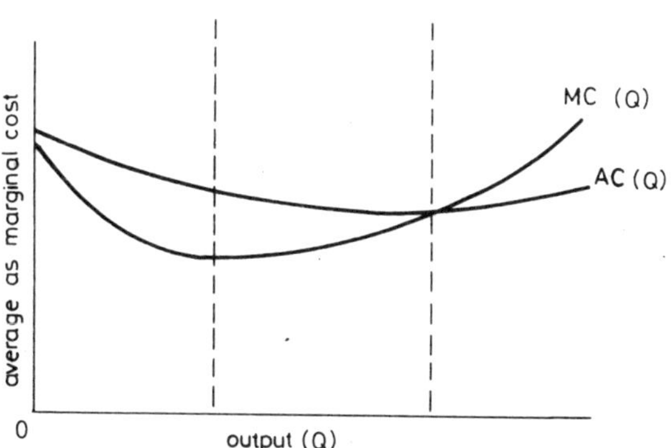

Fig. 5.3   *Corresponding average and marginal-cost functions*
Vertical broken lines correspond to those of Fig. 5.2

the slope of the total-cost function itself, which represents marginal cost.

Where a single output is involved, situations of increasing, constant and decreasing returns to scale correspond to falling, constant and rising average-cost curves. The exact shape of the cost function is therefore of considerable interest, but is much disputed, and will inevitably vary from one situation to another. Although average-cost curves are generally assumed to be U-shaped (as in Fig. 5.3), some empirical evidence suggests they are more likely L-shaped, i.e. after initially falling they remain about constant. For telecommunications, the available evidence seems to suggest, as we shall see, that average-cost functions are L-shaped for long-distance transmission; for subscriber exchanges there is less agreement.

5.4   Transmission cost functions

A simple numerical example may be helpful. In a study of Illinois Bell Telephone (Littlechild, 1970*d*) it was found that to set up a pole-wire transmission system over a 500 mile distance would cost $2 M for land, installation and power, plus $500 per circuit, up to a capacity of 250 circuits. Beyond that capacity, additional identical systems would have to be installed. Letting Q denote number of circuits, the cost functions are as follows (in dollars).

$$C(Q) = \begin{cases} 0 & Q = 0 \\ 2\,000\,000 + 5000Q & 0 < Q \leqslant 250 \\ 4\,000\,000 + 5000Q & 250 < Q \leqslant 500 \end{cases} \qquad (5.4)$$

$$AC(Q) = \begin{cases} 0 & Q = 0 \\ \dfrac{2\,000\,000}{Q} + 5000 & 0 < Q \leqslant 250 \\ \dfrac{4\,000\,000}{Q} + 5000 & 250 < Q < 500 \end{cases} \qquad (5.5)$$

$$MC(Q) = \begin{cases} 2\,000\,000 & Q = 0, Q = 250 \\ 5000 & 0 < Q < 249, 250 < Q < 500 \end{cases} \qquad (5.6)$$

The extensions to outputs over 500 circuits are obvious. These cost functions are illustrated in Figs. 5.4 and 5.5.

This is a convenient place to introduce some further terminology. The costs of $2 M for land, installation, power, etc. are sometimes referred to as *overhead* or *common costs* with respect to the number of circuits. The terms *fixed cost* and *variable cost* are sometimes used to refer, respectively, to the $2 M and the $5000. This usage should be treated with care. Certainly the 'fixed' cost here does not vary with the number of circuits within the capacity of the system, but it *does* vary accoring to whether a system is installed at all, and thereafter with the number of systems. At the moment of planning, *all* costs are variable. On the other hand, the 'variable' cost, which varies with number of circuits, does *not* vary with, say, the volume of traffic actually using those circuits, nor (let us assume) with the length of time for which the circuit is installed. One must, therefore, be careful to specify with respect to what, or in what sense, a cost is fixed or variable. The element of $5000 is often referred to as the marginal cost; this is almost always true, but not at outputs of 0, 250 and 500 circuits where one more circuit would necessitate an entire additional system. American

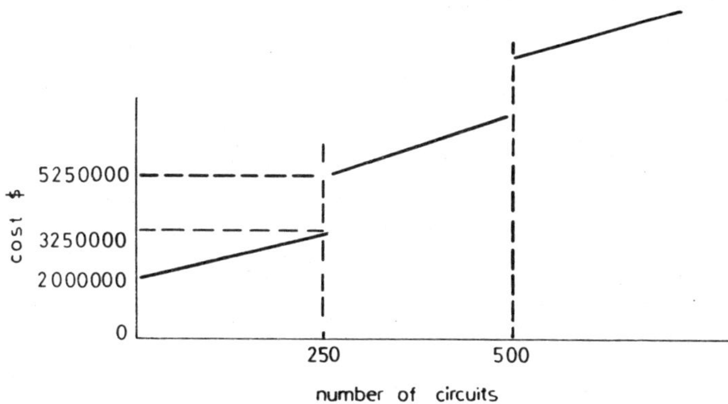

Fig. 5.4 *Total-cost function for pole-wire system*

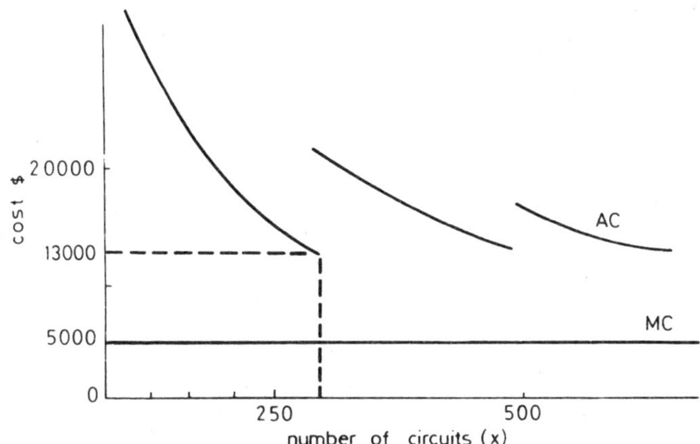

Fig. 5.5 *Average- and marginal-cost functions for pole-wire system*

engineers use the terms *starting-up cost* or *set-up* to denote the cost of operating a system of minimum viable size—in effect, marginal cost at zero output.

Fig. 5.6 reproduces the average-cost function for TD-2 radio-relay systems which was presented in testimony before the FCC by Albert Froggatt, (1966), a Vice-President of AT & T. The corresponding total cost function is

$$C(Q) = \begin{cases} 0 & Q = 0 \\ 15\,800 & 0 < Q \leqslant 600 \\ 16\,700 & 600 < Q \leqslant 1200 \\ 17\,600 & 1200 < Q \leqslant 1800 \text{ etc.} \end{cases} \tag{5.7}$$

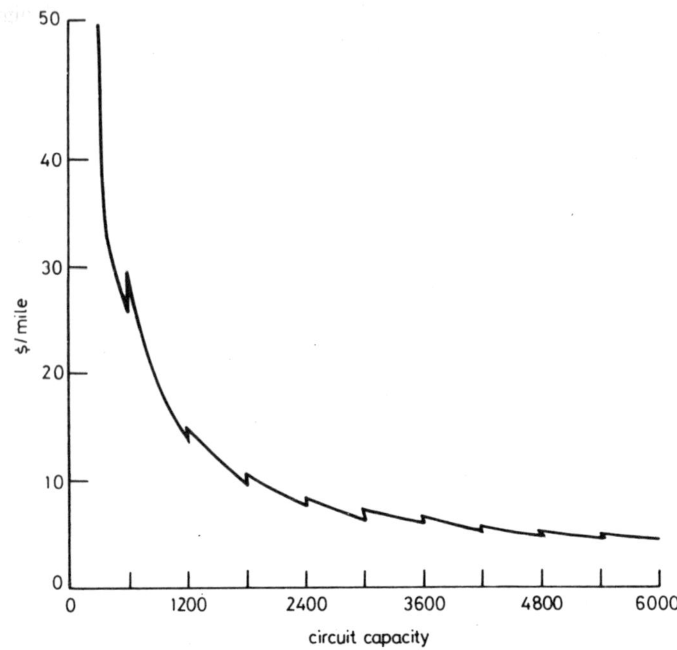

Fig. 5.6   *Illustrative average-cost function for TD-2 radio-relay system*
Book costs per voice circuit route mile [Source: Froggatt (1966),
chart 2]

The average-cost curve is quite clearly L-shaped. It exhibits a substantial
and more or less continuous fall from $26 per circuit route mile at 600
circuits (and even higher costs for lower volumes) down to $6·50 at
3000 circuits. The reason for this is that the common costs of towers,
building and access roads, antennas power and maintenance equipment,
which total $14 000, are being spread more thinly. At large capacities,
however, this element is not so dominant, and even a further doubling
of capacity to 6000 circuits reduces average cost only to $4. Thus in
1966 there seem to have been substantial economies of scale in radio-
relay systems up to about 3000 circuits, but only very slight economies
of scale thereafter.

Data for a 12-tube coaxial cable presented in the same testimony
again suggest substantial economies of scale which are still noticeable at
9000 circuits capacity. Mantell (1975a) refers to calculations based on
US transmission links which 'show that a size of one-tenth the average
1974 configuration would require a cost of 72·6 per cent greater, while
a system five times as large would diminish costs to 67·8 per cent'.

5.5   Incremental cost

Suppose, during the course of its future planning deliberations, that a telephone company considers the possibility of installing a radio-relay system with a capacity of 1000 circuits. It will want to know the total cost involved and probably the average cost per circuit. It will also want to consider possible alterations to this plan. Marginal cost reveals what the cost of one extra circuit will be—but how useful is that information? Is it not more likely that a more sizable increase (or decrease) would be envisaged—say, 250 or 500 circuits? Moreover, insofar as marginal cost is almost everywhere zero for this system, it can be positively misleading. It would be more useful and less misleading to know that the cost of an increment of either 250 or 500 circuits was the same (at $900 per mile) and that the average incremental costs per circuit-mile were $3·60 and $1·80, respectively.

    This can be represented formally. Let the *total incremental cost* of adding an increment Δ to an initial output Q be written

$$IC(Q, \Delta) = C(Q + \Delta) - C(Q) \qquad (5.8)$$

We may define also *average incremental cost* by

$$AIC(Q, \Delta) = \frac{IC(Q, \Delta)}{\Delta} = \frac{C(Q + \Delta) - C(Q)}{\Delta} \qquad (5.9)$$

The initial level of output is not to be regarded as something already existing, or produced earlier in time, but rather as a reference level, or plan, from which divergences may conveniently be assessed. The increment Δ could be positive or negative.

    The total cost of output Q can be thought of as the cost of an increment Q to an initial output zero, since

$$IC(0, Q) = C(Q) - C(0) = C(Q) \qquad (5.10)$$

where $C(0) = 0$ because it costs nothing to produce zero output. What were previously called average and marginal cost now appear as special cases of average incremental cost. The average cost of output Q is actually the average cost of an increment Q to an initial output of zero, for

$$AIC(0, Q) = \frac{C(Q) - C(0)}{Q} = AC(Q) \qquad (5.11)$$

Marginal cost at output Q is actually average incremental cost at output Q with an increment of one unit, for

$$AIC(Q, 1) = \frac{C(Q+1) - C(Q)}{1} = MC(Q) \qquad (5.12a)$$

or, for continuously differentiable functions, the limit as the size of the increment tends to zero:

$$\lim_{\Delta \to 0} AIC(Q, \Delta) = \frac{C(Q+\Delta) - C(Q)}{\Delta} = \frac{dC(Q)}{dQ} = MC(Q) \qquad (5.12b)$$

Average incremental cost is thus useful in practical situations, where cost functions are seldom continuously differentiable, and where calculation invariably proceeds in terms of discrete increments. Substantial and frequent indivisibilities render marginal cost quite erratic; averaging over a larger increment may be more satisfactory, even where marginal-cost pricing is the prescribed rule (see chapters 9 and 10). Incremental cost has the additional merit that it forces one to be precise about the initial level of output and size of increment envisaged, for incremental cost may vary quite considerably with both these parameters. In recent years, economists have tended to focus on marginal cost as the most important determinant of an optimal decision; in general, it is necessary to examine and evaluate *many* possible increments in output, including the increment which is total output (i.e. whether to produce at all). Average- and marginal-cost curves are frequently not sufficiently flexible concepts for this purpose.

5.6   Envelope cost functions

We have so far discussed the cost function corresponding to different types of transmission systems, notably pole wire and radio relay. In some circumstances the telephone administration may wish to restrict consideration to one particular system—for example, in the realm of switching, it may earlier have made a policy decision to stick with Strowger, or not to consider crossbar, or to install electronic exchanges. If this is the case, then the relevant cost function will be the one corresponding to the chosen system. If no such prior decision has been made—if, indeed, the task is precisely to choose a system—it is necessary to consider not only how the cost of each system varies with output, but also which system is least costly. It may well be that one system is least costly for one range of output and another system for a different range of output.

Suppose that the cost functions for pole-wire and radio-relay systems are as given in Sections 5.4. The common cost of the radio-relay system is much higher, but the cost per additional circuit is lower. It may be verified that for up to 500 circuits it is cheaper to install one or two pole-wire systems, but above 500 circuits radio-relay is cheaper. Of course, there are other technologies to consider, such as coaxial cables or satellites, which may be cheaper in the ranges 200–500 circuits or above 6000 circuits. In practice, it is also necessary to take into account the growth in traffic over time (Morgan, 1976). The example is intended for illustrative purposes only. The main point being made is that when a range of alternative systems is available, the relevant cost function is the *envelope cost function* obtained by taking the system with lowest cost at each level of output. Fig. 5.7 shows the envelope average-cost function based on the two systems discussed.

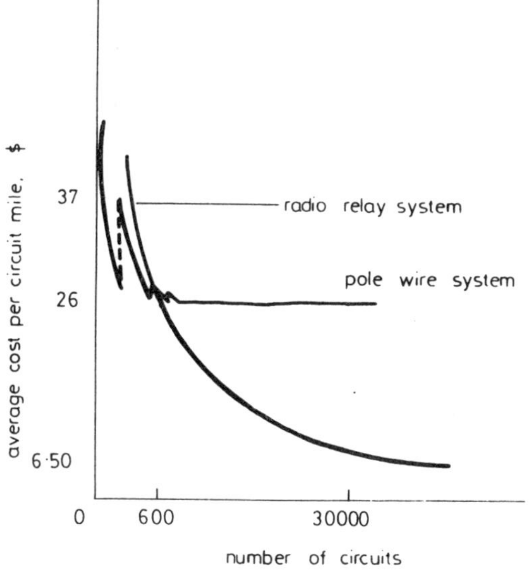

Fig. 5.7   *Illustrative envelope average-cost function for pole-wire and radio-relay systems*
Envelope curve shown by heavy line

We may note here for future reference a property of envelope curves which has been much utilised in discssions of marginal-cost pricing. Assume that there is a continuous range of technologies, each of which is the cheapest for some level of output. Since we are dealing with telecommunications, where economies of scale seem to be dominant, we may assume without loss of generality that each average-cost curve is broadly L-shaped. So, too, will be the envelope average-cost curve. Fig. 5.8 shows the envelope curve with typical component cost curves illustrated at three points.

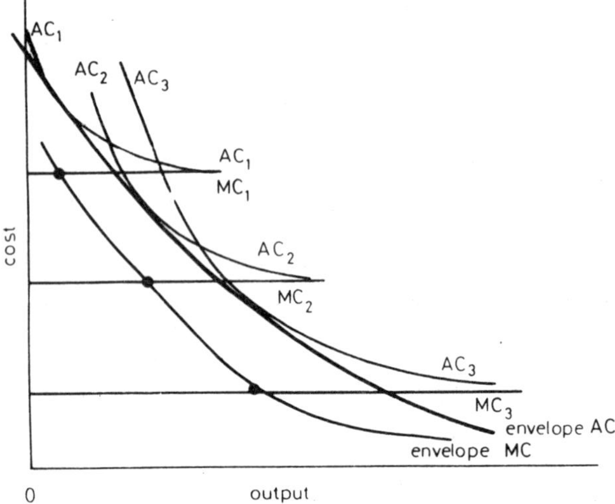

Fig. 5.8   *Properties of the envelope-cost function*

From the envelope average-cost curve we may derive an envelope marginal-cost curve. It shows the cost of one additional unit of output, on the assumption that the technology can, if appropriate, be changed (remember, this is the planning stage). whereas an individual cost curve refers to a given technology. The question arises—what is the relation between the envelope marginal-cost curve and the component marginal-cost curves? The relationship is both simple and interesting. Marginal cost for each individual technology is equal to envelope marginal cost at precisely the level of output where the individual average cost curve is tangential to the envelope average cost curve. In other words, if the correct technology is chosen for any given level of output, i.e. that which provides minimum average cost, then marginal cost at that technology is equal to envelope marginal cost (Ferguson 1971, Ferguson and Gould, 1975).

It thus appears that there are many quite different cost functions to consider—one cost function for each available technology and one envelope cost function. Actually, one could develop other cost functions by taking the envelope corresponding to a subset of techniques. Which cost function is relevant for decision making? It depends on the range of alternatives which the decision maker is willing to consider. If he is as yet uncommitted, and places no restrictions in his choice, he will use the envelope cost function. If he is committed to Strowger (or to submarine cables), he will use the Strowger (or submarine cable) cost function. If he wishes to buy British, but is otherwise uncommitted, he will use the envelope of the cost functions corresponding to British equipment, and so on. The less committed he is, and the fewer alternatives he rules out, the lower will be his envelope cost function.

5.7   Multiple outputs and inputs

Outputs and inputs need not possess only a single dimension. Consequently, a cost function may refer to a whole vector of outputs or inputs. For example, the cost of setting up a local exchange will depend partly on the number Q_1 of subscriber lines provided and partly on the volume Q_2 of busy-hour traffic. The following total-cost function has been used to characterise a typical Illionois Bell central office (Littlechild, 1970d):

$$C(Q_1, Q_2) = \begin{cases} 0 & Q_1 = Q_2 = 0 \\ 1\,007\,000 + 23Q_1 + 27Q_2 & \text{otherwise} \end{cases} \tag{5.13}$$

where busy-hour traffic Q_2 is measured in ccs per hour and cost is measured in dollars. The elements \$23 and \$27 are said to be *separable* or *attributable* costs; the element \$1 007 000 is common to both outputs and is said to be a *joint* or *common* cost.

The terms joint and common cost are often used interchangeably; some authors reserve the term common to costs which are independent of the level of output, once the decision has been made to produce *some* output, as in eqn. 5.13, while joint refers to costs which are directly or indirectly variable with output. An example of this latter usage is the cost of providing capacity at different times of day. Suppose outputs in each of the 24 hours of the day are regarded as different products. The cost of providing capacity for outputs Q_1, $Q_2, \ldots, Q_{24}$ is equal to the cost of providing capacity in the busy-hour, hence

$$C(Q_1, Q_2, \ldots, Q_{24}) = c \max [Q_1, Q_2, \ldots, Q_{24}] \tag{5.14}$$

or, alternatively,

$$C(Q_1, Q_2, \ldots, Q_{24}) = \min cy \tag{5.15}$$

subject to $y \geqslant Q_i, i = 1, 2, \ldots, 24$, where c is the joint cost per unit of busy-hour capacity.

For empirical purposes, it may be useful to relate total costs to a variety of observable inputs. Waverman (1975) has estimated the following total-cost function for private microwave systems in the USA, using data on 141 systems with greatly varying grades of service, distance and capacity:

$$\log C = -0 \cdot 94 + 0 \cdot 23 \log S + 0 \cdot 45 \log U + 0 \cdot 83 \log H + 0 \cdot 43 \log G$$

$$(5.16)$$

$$\bar{R}^2 = 0 \cdot 92$$

where

    C = total cost

    S = number of circuits available (capacity)

    U = number of operating circuits

    H = number of hops (number of stations minus one)

    G = dummy, with 1 for toll-grade system, 0 otherwise

All the coefficients are statistically significant, and together they provide a very good explanation of total cost. Doubling circuit capacity increases total cost by 23%, doubling circuits used increases total cost by 45% and doubling the number of hops increases total cost by 83%. The author concludes that his results confirm the existence of scale economies up to a capacity of 1000 circuits. This is a somewhat lower level than the theoretical calculations of Froggatt (1966) had suggested.

Failure to take account of the multiple dimensions of output has invalidated many studies of the costs of local telephone service. It used to be argued that telephone exchanges were characterised by increasing average costs per customer for local service (Clemens, 1950, and references cited by Rose, 1950) and two later empirical studies appeared to support this view (Meyer, 1961, and Bowers and Lovejoy, 1965). Rose himself argued for an inverse relationship between average cost and number of subscribers, i.e. economies of scale, and Mantell (1975a) reports calculations that a tenfold increase in subscriber density (number of telephones per unit area) would lead to a reduction in average cost varying from 15% for large exchanges (with 2000 to 20 000 terminals per square mile) up to 47% for smaller exchanges (with 100 to 1000 terminals per square mile). The difficulty with the majority of these and other studies, as Alleman (1977, Chap. 5) points out in a very useful survey, is that they fail to distinguish the various dimensions of output. The cost of providing local telephone service will surely depend on the number of subscribers, the volume of traffic and the total area served. It will also depend on the geographical character of the area (whether a Chicago suburb or Texas plains or tropical mountainous jungle) and on the calling patterns of the subscribers (whether predominantly business or residential, whether high or low income etc,). To look at one of these variables alone, such as number of subscribers, will give very misleading results. This is an important area for further research, since a number of current policy decisions could depend on the answers obtained. For example, if the entry of new competitors into telecommunications in the USA will have the effect of bringing prices and costs closer into line, it becomes important to estimate what costs are to predict the effect on charges for local telephone service (Allemann, 1976, and references therein; some casual estimates of local exchange costs are made by Littlechild, 1970*d*, Littlechild and Rousseau, 1975 and B.M. Mitchell, 1978*b*).

5.8   The allocation of joint costs

Joint and common costs abound in telecommunications, and probably in most industries. For example, a radio-relay system may be used to transmit microwave signals for private line telephone, telephotograph and teletypewriter services, TV programme transmission etc. (Froggatt, 1966, p. 5). Administrators and regulators, and even some businessmen and accountants, have a great desire to *allocate* overhead, common or joint costs between the various outputs which they make possible. Morgan (1976) quite rightly warns against using such allocated cost figures in investment appraisals. Sometimes this has to be done for tax or bookkeeping purposes, as when depreciation is applied to the capital cost of a piece of equipment, thereby allocating the common cost between the outputs in succeeding years. Another example is the 'separations procedures' used by US telephone companies to allocate plant between interstate and intrastate usage, which is then used as a basis for dividing up telephone revenues between Long Lines, the Bell operating companies and the independent companies. (Gabel, 1967 D.C. Mitchell, 1978). Sometimes the purpose is to calculate 'fair' costs for the purpose of setting or appraising prices. A prime example of this was the Seven Way Cost Study of 1965, in which the FCC required Bell Telephone to apportion all its costs between its seven main classes of service (Kahn, 1970–1971, p. 71, Bolter, 1978).

Pricing and regulation are not our concern in this chapter, but it is appropriate to emphasise that the result of such an arbitrary allocation procedure is inevitably a figure of 'cost' which is not equal to the cost of anything. Thus, in the case of the Illinois Bell central office with the cost function of eqn. 5.13, the common cost of $1 007 000 must be taken into account in deciding whether or not to build the office.

However, once the decision to build has been taken, the common cost is irrelevant in calculating the optimal number of subscriber lines and volume of busy-hour traffic. It would be quite misleading to suggest that the average cost of a subscriber line is $23 plus a specified share of $1 007 000 pertaining to lines alone. Of course, for the office as a whole to pay its way, the revenues from subscriber line rentals plus telephone calls must cover the total cost of the office, including the common cost, but in setting the prices of these services it does not help to allocate the common costs among the different kinds of output.

Similar problems can arise in long-term planning. We may write the outputs $Q_1, Q_2, \ldots, Q_n$ in a cost function to denote output in years (or months) $1, 2, \ldots, n$ up to whatever planning 'horizon' (terminal date) seems appropriate. The criteria for choosing outputs, methods of production and levels of investment, and in particular the weighting of future costs against present costs, will be dealt with in chapter 7. Here we are interested in how (if at all) one might answer a question such as—what is the cost of output in some specified year t? We shall first need to rephrase the question in terms of incremental cost—what is the change in total cost caused by making a specified change in the planned output? This procedure avoids all the awkward (and ultimately unanswerable because ill-founded) questions about the appropriate allocation of costs to the multiple outputs which they make possible.

A simple numerical example will illustrate this. Suppose a telephone company is considering installing a manual exchange in a rural community. The cost of the capital equipment, which has a technical life of 25 years, is estimated at £500 000; wages to operate and service the exchange (including maintenance and repairs) are estimated at £250 per week, or £13 000 per year. In answer to the questions—what is the cost of this service in the first year? in the second year? and so on — some would suggest allocating the capital cost equally between the 25 years, i.e. £20 000 per year; hence the cost in each year is £13 000 + £20 000 = £33 000. This is a well-intentioned answer to a naive question. Costs, as emphasised earlier, are associated only with decisions between alternatives. What are the relevant decisions here? Presumably, whether or not to provide service in the specified years, and, in particular, when to begin service. A convenient way to analyse these decisions is to examine the incremental cost of providing service, i.e. to compare the cost of the original plan with that of a modified plan in which service is not provided in a specified year.

What is saved by deferring the date of installation from the first to the second year? Certainly, the operating and service cost of £13 000. Also, the interest that can be earned for a year on the £250 000 that would otherwise be tied up (this aspect is dealt with in chapter 7, which considers investment appraisal). Perhaps, also, if purchase is postponed for a year, a cheaper or more efficient exchange can be purchased, depending on the rate of technical progress. It would be remarkable if the incremental cost of the first year's service, given by the total of these elements, were exactly equal to £33 000. The inadequacy of the allocated cost figure is even more striking if we consider that the cost saved by not providing service in the second year, assuming that service is to be provided in the first and subsequent periods, is merely that of maintenance, i.e. £13 000.

These calculations in themselves do not provide any basis for deciding whether or when service should be provided; they tell us only the savings from not providing service in a specified year. To make the correct decision one must introduce the benefits of providing service. This topic must wait until later chapters.

5.9   The passage of time

We have hitherto assumed that the telecommunications administration starts with a clean slate. It owns money but no other resources. All the planning calculations we have so far discussed, concerning investments and outputs, refer to future events as conjectured from the present.

We must now face the fact that time moves on. Decisions are taken which have consequences for the business. Assets are acquired and sold, commitments are made and completed. Some factors of production have relatively short lives, such as office supplies, while others have very long lives, such as outside plant. Some factors, notably staff, are hired on a weekly, monthly or annual basis; other factors are leased for a fixed period of time, perhaps land for office buildings; yet other factors, such as outside plant or exchange equipment, are typically bought outright and kept until their economic lives are exhausted.

If the administration had perfect knowledge of the future, it could conceivably draw up one single plan for ever, which would not need revising. In practice, this is impossible. Mistakes are always made, opportunities are overlooked, surprises and new developments occur. It is, therefore, necessary for the administration to revise its plans from time to time. At such moments, the administration typically finds itself with (i) a collection of plant and equipment, which it owns, of various kinds and vintages, and with various lengths of useful life remaining, (ii) a set of commitments to suppliers, employees, customers and government and (iii) various opportunities to extend or contract its business, to hire or train new staff, to introduce technological developments etc.

The principles we have already outlined remain perfectly valid. However, we need to take into account the treatment of existing assets. The fundamental point here is that once costs have been incurred, they are no longer relevant for present and future decisions; they are said to be *sunk* costs. For example, if a company takes out a 12-month lease on a building, the money paid for the lease, even if it is spread over the year in monthly instalments, does not affect the costs of subsequent decisions during that year concerning the use to which the building is put. The company is already committed to these payments; they do not depend on decisions about the use to which the building will be put, and should be ignored in making those decisions. The payment for the lease did *at one time* constitute a cost, when the decision was made to take out the lease, but now this cost is *sunk*. Of course, when the time comes to decide about renewing the lease, the prospective payments to be paid in the forthcoming year will constitute a valid cost of renewal.

Although the price paid to obtain an asset is henceforth irrelevant, the fact of the asset's existence, and the opportunities for use which it presents, most decidedly *are* important for planning purposes. Two other concepts of cost are frequently used in this connection.

A company may own and use assets which it would otherwise have to hire from another supplier. For example, the owner of a small business may use his own garage for storage, instead of renting space elsewhere. The owner will typically not pay himself a rent for the garage, but he should recognise that part of his gross profit is, in effect, a payment for the use of the garage. The amount which he would otherwise have to pay is called the *implicit cost* of using his own garage. Failure to recognise implicit costs can lead to misleading estimates of the true profitability of a business. For example, it is often suggested that if proprietors of small corner shops took into account the value of their own time, then the small bookkeeping profits they appear to make would, in fact, be recognised as losses (but, in turn, the prevalence of such corner shops suggests that the proprietors derive some satisfaction over and above the money revenues which the accounts record).

An even more fundamental notion is that of *opportunity cost* (Alchian, 1968). Any asset may be used in a variety of ways; to use it in any one way precludes using it another way. The sacrificed profit on revenue which could have been earned in the next-best usage constitutes the opportunity cost of using an asset. To say that an asset should be used in its most profitable way is equivalent to saying that the revenue it generates will cover its opportunity cost. Note especially that opportunity cost is a forward-looking concept; it depends on conjectures about the future. This is in contrast to the backward-looking concept of historical cost and sunk cost.

A simple example will illustrate. Suppose a company has signed a 12-month lease on a temporary warehouse, but after one month realises that it has overestimated its storage requirements. It would be fallacious to argue as follows: the revised value of storage facilities is £A which is less than the rent £B which has to be paid for the remaining 11 months, therefore, it is too expensive to use the warehouse. Certainly, the original decision to rent the warehouse was a mistake, but the outstanding commitment of £B is now irrelevant to future decisions. What counts are the alternative available uses of the warehouse. One possibility is to negotiate an early end to the lease in return for a rebate of £C on the total originally due; a second possibility is to sublease to another company for £D; a third possibility is store some other goods in the warehouse at a saving of £E in storage costs elsewhere. The company must now ask whether the revised value of the warehouse for original purpose (£A) exceeds the opportunity cost of using it in this way, which is the maximum of £C, £D and £E. If so, continue, if not, cancel or sublease or change the usage accordingly.

The concept of opportunity cost is of great importance for pricing. A telecommunications administration may have installed large amounts of crossbar equipment before electronic switching became viable. The decisions about whether and how fast to change to an electronic system should be taken in the light of future costs and benefits; the cost of originally installing the crossbar system is now sunk and irrelevant (except insofar as it imposes future constraints on revenue to cover interest payments). Similarly, investment in submarine cables or satellites may have been made in the expectation of a high traffic rate at high prices. If new competitors and new technology render these original expectations invalid, the correct pricing strategy should take into account future responses by competitors, customers and government, but not the sunk costs of the original investments. We turn to such questions of investment and pricing in the next two chapters.

5.10 A note on the short run and the long run

One attempt to incorporate time into economic analysis uses a distinction between the *short-run* and the *long-run*. In the short-run, it is assumed that certain specified factors of production are fixed (typically, capital equipment) and others are freely variable (typically, labour). The *short-run cost function* is the cost function which results from varying, in the least cost manner possible, those inputs which can be varied, the remainder being fixed. In the long run it is assumed that all factors are freely variable, and the *long-run cost function* is obtained by optimally combining all inputs. Evidently the short-run cost function will always lie above the long-run cost function — in fact, the relationship between the two is precisely that the long-run cost function is the envelope of all the short-run cost functions, as described in the previous Section.

Nevertheless, there are certain limitations about the concepts of the short-run and the long-run which cause us not to favour them (Alchian, 1959, Turvey 1969). First, in practice, no factor is completely fixed; it is variable at a price, and whether that price is worth paying will surely depend on the benefits of doing so. Thus one might conceive of a whole range of short-run functions corresponding to any given situation, depending on the assumptions one makes about fixity of different factors. Second, the long-run cost function does not embody the set of alternatives available to the firm at any time, but is merely a set of alternatives which would be available if things were different (specifically, if the firm had no existing commitments). Finally, this approach does not provide a mechanism for analysing the process by which a firm actually changes its fixed factors, depending on the cost of change, the time it will take, the benefits of doing so etc. In other words, the notion of short-run and long-run is perhaps not the most useful device for analysing the problems of change over time. In the present and next chapters, we suggest alternative procedures, in which time is incorporated explicitly. The distinction between the short-run and long-run is no longer needed.

5.11 Uncertainty and information

We have hitherto looked at the problem of planning over time from the viewpoint of a single moment at (or just before) the beginning of the planning period. In practice, an organisation never makes permanent plans for a long period because to do so would be an extremely complex business and it would prevent taking advantage of new knowledge acquired with the passage of time. Plans are, therefore, revised whenever it seems worth while to do so. New opportunities may be perceived (e.g. new products, customers or methods of production etc.)

or it may be realised that existing plans are based on erroneous assumptions (e.g. about demand, prices, availability of resources etc.).

This new information may well change costs. Obviously, if wages increase, the cost of manning a telephone exchange goes up. Less obviously, if the demand for circuits by television companies goes up, so too does the cost of providing telephone calls on certain routes, simply because the opportunity cost of using those circuits for telephone calls in now higher than before. We shall refer to these matters again in Section 7.2 and chapter 14.

Chapter 6

The telephone system in Costa Rica: a case study

6.1   Introduction

To illustrate the notions of demand and cost discussed in the previous chapters, it will be useful to show how some of these ideas may be applied in quite a simple way to an actual telephone system. The following case study is based on a report prepared by the author (Littlchild, 1973a) acting as a consultant to the Public Utilities Division of the International Bank for Reconstruction and Development (the 'World Bank'). The Bank was then lending for telecommunications projects at the rate of over one hundred million US dollars per year. It had considerable expertise in technical and financial matters, but wished to gain a better knowledge of the economic aspects of telecommunications. For several reasons the telephone system in Costa Rica appeared to be a suitable choice; happily, the Costa Rican telecommunications organisation was prepared to co-operate. The report was prepared after a visit in March 1973, but most of the cost data refer to the first phase of construction, which was completed in December 1970. An extremely thorough economic analysis of the Costa Rica system has subsequently been provided by Schkolnick (1976). The telecommunications organisation itself has since produced several excellent and detailed feasibility studies and other papers (ICE, 1973, Bonilla, 1977, Cañas, 1977).

6.2   The telephone system in Costa Rica

Costa Rica covers an area of 51 000 square km and is located in the narrow southern part of Central America between Nicaragua and Panama. It is a mountainous country except along the Caribbean and Pacific Coasts. The population, now about 1·8 million, has doubled

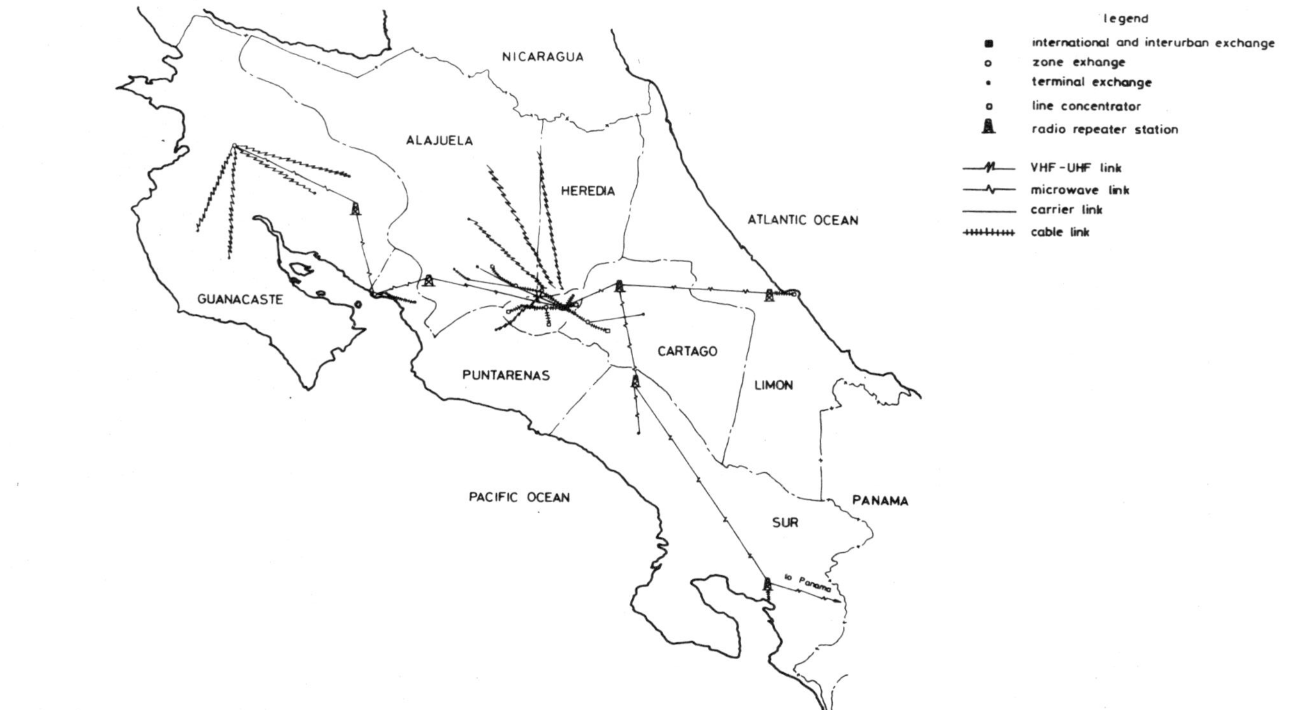

Fig. 6.1   *National Telephone System of Costa Rica at end of 1970*

over the past 20 years. Two-thirds of the people live in the central highlands, where the capital San José is located and where the climate is mild throughout the year, in contrast to the tropical climate of the coastal plains.

Costa Rica achieved remarkable economic growth in the 1960s, with gross domestic product increasing by nearly 7% annually. Because of the rapid population growth, however, national income per capita increased by little over 3% per annum.

The unit of currency is the colon (¢). At the official rate of exchange in 1973 there were 6·65 colones to one US dollar, but on the free market the exchange rate was ¢8·50 to $1.

Both telecommunications and power are provided by the Institute Costarricense de Electricidad (ICE), which is owned by the government. Most of the telecommunications system existing in 1970 was established under ICE's Stage 1 expansion plan, financed by a World Bank loan. Automatic telephone service was introduced in 1966, based on crossbar equipment. The use of a pulse system to meter and charge for calls means that very flexible pricing systems are possible; on the other hand, information about number, destination and duration of calls is not easily available.

The system is designed according to the familiar star pattern (Fig. 6.1). At the centre are the six metropolitan area exchanges of San José and an interurban switching centre at San Pedro, also located in San José. Links radiate to exchanges in the six other provincial capitals, and from these to the exchanges in the smaller towns within

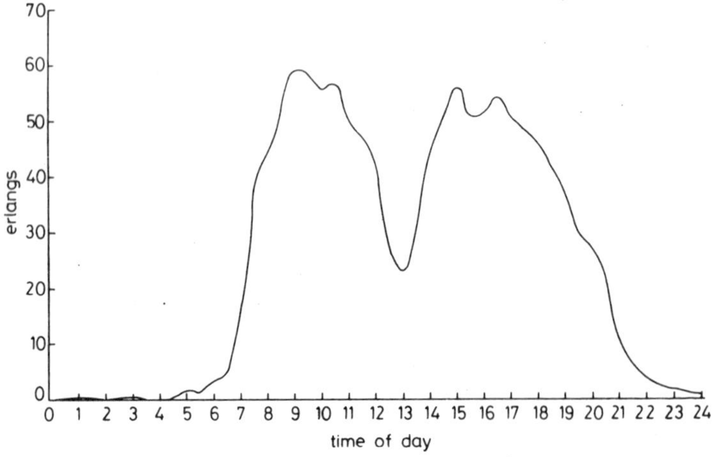

Fig. 6.2 *Weekday 24-hour load curve for link between Alajuela Exchange and Interurban Switching Centre (a Thursday)*

those provinces. Finally, several rural villages are each connected by one or more circuits to their local towns.

In 1972 there were some 60 000 telephone lines in Costa Rica, giving a telephone density of 33 per thousand population. This compared with an average of 28 per thousand for central American and Caribbean countries. The bulk of the telephones, as with the population, is in the metropolitan area, where the density of lines is over twice as great as in the country as a whole. Indeed, we shall see that over three quarters of the lines are in the metropolitan area and that the majority of telephone calls are made either to or within the metropolitan area.

The load curve is highly peaked, as Fig. 6.2 illustrates for one link. Severe congestion occurred in several exchanges, notably the Interurban Switching Centre (ARM) in San Pedro.

6.3   Income statement and balance sheet

Another picture of the telephone system may be obtained from the Income Statement for 1970 and the Balance Sheet at the end of that year (Table 6.1). Income totals ¢39 M of which ¢11 M, or nearly one-third, is profit. Plant in operation accounts for about two-thirds of the system's assets of ¢191 M, and long-term borrowing (from the government) accounts for about two-thirds of its capital and liabilities.

Table 6.1

Income statement (year ending 31 December 1970)

		¢M
Income from telephone services		33
Other income (including international)		6
Total income		39
Less Wages	7	
Depreciation	6	
Other expenses	8	21
Net operating revenue		18
Less Finance charges		7
Profit for the year		¢11 M

Table 6.1 *(Continued)*

Balance sheet (as at 31 December 1970)

Assets   ¢M

Plant in operation	126		
less cumulative depreciation	18	108	
Under construction		11	
Other fixed assets	8		
less cumulative depreciation	3	5	
Total fixed assets			124
Inventories			35
Other current assets			14
Other assets			12
Other investments			6
			¢191 M

Liabilities

Long term	131	
Short term	13	
Other	5	149

Capital

Telephone subscribers	13	
Reserves	11	
Other	6	
Net profit for year	12	42
Total liabilities plus capital		¢191 M

6.4   The telephone tariff

The following tariff had been in operation since 1965:

(a)   Installation fee

A nonreturnable installation fee is payable according to the following scale. Extra charges may be made for distant subscribers:

Type of service	Metropolitan area	Provincial Capitals	Other
Commercial	¢	¢	¢
(including PBX)	444	372	300
Residential	336	288	240
Shared	228	192	156

The tariff for each private exchange (PBX) line is as for a commercial line. A refundable deposit of ¢50 per line is required as a guarantee of payment. A change of address costs ¢75 per line.

(b)   Rental
The following monthly rentals apply:

Type of service	Monthly rental
	¢
Commercial	37
Residential	28
Party line	19

Note that the metropolitan-area installation fee for each type of sub-scriber is equal to one year's rental for that type of subscriber.

(c)   Call charges
Calls are recorded in terms of 'pulses', as set out below. Each line is allowed 100 free pulses per month (50 for shared lines) and each subsequent excess pulse is charged at the rate of ¢0·10:

Local calls: One pulse every 5 mins (local calls are those within an exchange or within the metropolitan area).

Long-distance calls: The interval between pulses depends on distance. A rough calculation suggests that pulse rate as a function of distance is approximately one-and-a-half pulses plus one pulse per 30 km.

For local calls from 6 p.m. to 8 a.m. and for long distance calls from 8 p.m. to 6 a.m. and on Sundays and public holidays, the interval between pulses is doubled. Thus, night-time rates are approximately half daytime rates.

(d)   Public telephones
A subscriber may have a public telephone installed at a fee of ¢666 (metropolitan area), ¢558 (provincial capital) or ¢450 (other locations). The monthly rental is ¢50. Pulses for local calls are charged at ¢0·25 each. For long-distance calls, the charge is ¢0·18 per pulse, but the interval between pulses is doubled. The subscriber has the right to 25% of the revenue (over ¢50) up to ¢300, and 15% thereafter.

6.5   Subscribers, calls and revenues

Records are available, month by month and exchange by exchange, of the number of lines, number of excess pulses and total number of pulses for each type of subscriber. The record for December 1970 shows that there were some 35 706 lines paying an average monthly rental of ¢30·57 and with an average excess usage of 431 pulses per month. Assuming December is typical, the general picture of annual revenue from telephones is therefore:

Rental:	35 706 lines @ ¢30·57 per month x 12	¢13·1 M
Excess charges:	35 706 lines x 431 pulses per month @ ¢0·10 x 12	¢18·5 M
	Annual income from telephones	¢31·6 M

These data thus account for over 95% of the ¢33 M income from telephone services shown on the Income statement. It appears that about 40% of total revenue comes from rental charges and about 60% from excess charges.

It is of interest to provide a more detailed breakdown of the location and type of subscribers, their numbers of pulses etc.:

(a)   Location and type of subscribers (Table 6.2)
Over three-quarters of the 35 706 telephone lines were situated in the metropolitan area, and most of the remainder in the provincial capitals. Nearly two-thirds of the lines were residential, and one-third were commercial or PBX. Thus, almost half the lines belong to residential subscribers in the metropolitan area, and one-quarter to commercial or PBX subscribers there.

Table 6.2   *Percentage distribution of lines (December 1970)*

	Metropolitan area	Provincial Capitals	Other	Total
PBX	5	0·4	0·1	5
Commercial	20	6	2	28
Residential	48	11	3	62
Shared	4	1	0·1	5
Total	77	18	5	100%

Note: 100% of telephone lines is 35 706

(b)   Lines per head of population (Table 6.3)
Not only is the number of telephone lines higher in the metropolitan area, but so is the density of telephone lines per head of population. In those areas with telephone exchanges, the average density throughout the country was 44 lines per thousand population. The density in the metropolitan area was half as high again, whereas in rural areas it was only 11 lines per thousand population.

Table 6.3   *Telephone lines per head of population (December 1970)*

	Metropolitan area	Provincial Capitals	Other	Total
Number of lines	27 548	6480	1678	35 706
Population	402 957	254 030	156 507	813 494
Lines per '000 population	68	26	11	44

(c)   Excess pulses (Table 6.4)
In December 1970, the average number of excess pulses per line per month was 431. Subscribers outside the metropolitan area used twice as many excess pulses as metropolitan area subscribers. Investigations by the telephone administration suggest that the majority of calls are made to or within the metropolitan area. The fact that long-distance calls require many more pulses probably explains the difference in excess pulses. PBX lines used over three times the average number of excess pulses, while residential and shared lines used about half the average number.

Table 6.4   *Ratio of average number of excess pulses, by type and location of subscriber, to overall average (December 1970)*

	Metropolitan area	Provincial Capitals	Other	Average
PBX	3·0	6·5	8·0	3·3
Commercial	1·2	2·7	2·7	1·6
Residential	0·5	0·8	0·8	0·6
Shared	0·4	0·5	0·6	0·4
Average	0·8	1·5	1·7	1·0

Note: 1·0 represents overall average of 431 excess pulses per line

(d)   Telephone bills (Table 6.5)

From the same records we may calculate the average monthly telephone bill by type and location of subscriber, consisting of monthly rental plust monthly charge for excess pulses. The overall average bill is ₡74 per month (about ₡8·70 at the free market exchange rate). PBX lines typically pay over twice that amount, residential lines somewhat less.

Table 6.5   *Average monthly telephone bills — rental plus excess pulse charges (December 1970)*

	Metropolitan area	Provincial Capitals	Other	Average
	₡	₡	₡	₡
PBX	37+129=166	37+280=317	37+343=380	37+144=181
Commercial	37+53 = 90	37+117=154	37+116=153	37+71 =108
Residential	28+21 = 49	28+35 = 63	28+36 = 64	27+24 = 52
Shared	19+17 = 36	19+22 = 41	19+25 = 44	19+18 = 37
Average	31+36 = 67	31+76 = 98	31+72 =103	31+43 = 74

(e)   Annual revenue (Table 6.6)

The annual revenue from these rentals and excess pulse charges amounts to some £31·6 M, as indicated earlier. Two-thirds come from the metropolitan area and one quarter from the provincial capitals. Splitting it another way, commercial and residential customers each account for just over 40% of total revenue.

Table 6.6   *Sources of total revenue (December 1970)*

	Metropolitan area	Provincial capitals	Other	Total
PBX	11	2	0·4	13
Commercial	24	12	4	41
Residential	32	9	2	43
Shared	2	0·5	0·1	2
Total	69	24	7	100%

Note: 10% of total income for December, 1970, is ₡2 631 701, hence our estimate for the whole year is 12 times that, i.e. £31·6 M

6.6   Estimating demand functions

We wished to get a better understanding of how the number of connections (lines) and number of calls related to various explanatory
factors. In the time available, only a few rough calculations could be
performed. It was not possible to obtain information about individual
subscribers, but only about individual exchanges. There are several
other difficulties discussed as we proceed. The following three regression equations are presented, not so much for the value of the information they contain, as to illustrate some of the difficulties and dangers
of a superficial analysis of demand.

The first equation (6.1) regresses the total number of lines at each
exchange y_1 on the population served by the exchange x_1, the distance
in kilometres to San José x_2 and the distance in kilometres to the
provincial capital x_3. t statistics are given in parentheses below the
regression coefficients. They indicate the ratio of the coefficient to
its standard error, and hence measure the reliability of the coefficient.
Roughly speaking, with about 20 observations a t-statistic exceeding 2
in absolute value indicates a significant explanatory variable (Johnston,
1963).

$$\ln y_1 = -0\cdot27 + 0\cdot80 \ \ln x_1 - 0\cdot33 \ \ln x_2 - 0\cdot20 \ \ln x_3 \qquad (6.1)$$
$$\ \ \ \ \ (0\cdot2) \ \ (7\cdot5) \ \ \ \ \ \ \ \ (-5\cdot4) \ \ \ \ \ \ \ \ (-3\cdot7)$$

$$R^2 = 0\cdot94 \quad (22 \text{ observations})$$

This equation suggests that number of lines increases with the
population served by each exchange, but not quite proportionately. A
10% increase in population is associated with an 8% increase in number
of lines. It follows that number of lines *per head* is lower in exchanges
serving larger populations, i.e. in San José and the provincial capitals.
At first sight this result is surprising because it appears to contradict
our earlier calculations in Table 6.3. Presumably, the explanation is
that the two other variables, relating to distance, outweight the population effect. An exchange which is twice as far away from San José
as another exchange is likely to have one-third less lines; an exchange
which is twice as far away from its provincial capital is likely to have
20% fewer lines. We noted earlier that call prices are strongly related to
distance. It is possible that these negative coefficients reflect the
higher cost of making calls to San José from the outlying areas. In such
areas the net benefits of owning a telephone are lower than in the
capital. Hence, demand for connections is lower away from the national
or provincial capitals. By this argument, the negative regression co-

efficients can be interpreted as indications of price elasticities of demand for telephone connections.

One objection to this interpretation must immediately be noted . In many of the exchanges there is a shortage of capacity and a waiting list for lines. The number of lines we observe in those equations is a measure of *supply*, not demand. To this extent, the equation we have estimated reflects the provisioning policy of the telephone organisation rather than the demand of its customers.

Since the majority of telephone calls are made to the metropolitan area, it seems reasonable to approximate the number of calls per line by dividing the total number of pulses per time by the price (in pulses) of a call to San Jose of average duration. An average duration of 40 seconds was suggested to us, although it seems to be on the low side. Let y_2 and y_3 denote the average number of pulses per commercial and residential line, respectively; let p denote the price per call to San Jose; as before, let y_1 denote the total number of lines and x_3 denote to provincial capital. We estimated

$$y_2/p = 405 + 0 \cdot 15x_3 + 0 \cdot 014y_1 - 21 \cdot 91p$$
$$(11 \cdot 9) \ (0 \cdot 7) \quad (3 \cdot 0) \qquad (-3 \cdot 7) \qquad \qquad (6.2)$$
$$R^2 = 0 \cdot 68 \ (20 \text{ observations})$$

$$y_3/p = 194 + 0 \cdot 06x_3 + 0 \cdot 007y_1 - 12 \cdot 49p$$
$$(9 \cdot 9) \ (0 \cdot 5) \quad (2 \cdot 7) \qquad (-3 \cdot 7) \qquad \qquad (6.3)$$
$$R^2 = 0 \cdot 67 \quad (20 \text{ observations})$$

The distance x_3 to the provincial capital is not a significant determinant of the number of calls for either type of call, but the number of lines y_1 is significant. An extra 1000 lines will generate an additional 14 calls per month per commercial line, 7 calls per month per residential line. Evaluated at the mean points, the elasticity of demand for calls with respect to number of lines in the exchange is $0 \cdot 07$ for commercial lines, $0 \cdot 09$ for residential lines. These results suggest that larger calling areas do generate more traffic (cf. the discussion on externalities in chapter 12). However, it should be noted that the shortage of lines is most severe in the larger urban exchanges. It may be, then, that the smaller exchanges have been able to admit many subscribers with rather low demand for calls, whereas, in the larger exchanges, these subscribers have not yet been admitted. Since telephones may be transferred privately in Costa Rica, it is possible for a nonsubscriber with high demand to purchase a telephone line from an existing subscriber willing to sell, at a price which is mutually agreed between the two parties.

Price per call to San José is also a significant determinant of number of calls. If these coefficients can be taken at face value, they indicate price elasticities of demand, evaluated at the mean point, of -0.30 for commercial calls and -0.40 for residential calls.

These two sets of results are encouraging and suggestive, but the data are too aggregated and the econometric techniques too simple to give one much faith in the results. A more thorough study would give more reliable results, but against that one has to weigh the costs of carrying out such a study. It may well be that some simple calculations and regressions can be devised to serve as a useful complement to the routine forecasting carried out by the telephone organisation.

Table 6.7 *Stage-I construction expenses (up to 31 December 1970)*

Local system	Metropolitan area	ARM*	Provincial capital	Other places	Total
	¢ M	¢ M	¢ M	¢ M	¢ M
Ducts	6·4		2·7	0·2	9·4
Primary network	6·9		2·0		8·8
Secondary network	5·6		2·7	7·7	16·1
Installation of telephones	5·2		1·2	0·4	6·8
Installation of public telephones	0·3				0·3
Buildings	8·9		6·2	2·7	17·8
Switching equipment	20·2	2·6	6·4	3·5	32·7
Total local system	53·5	2·6	21·2	14·5	91·8

Transmission	Direct cost ¢ M	Overhead† ¢ M	Total cost ¢ M
Local transmission	0·7	0·2	0·9
Long-distance transmission	16·8	4·8	21·6
Total transmission	17·4	5·0	22·5

* Interurban Switching Centre at San Pedro

† Overhead appears to be calculated at the rate of 28·7% on direct costs. I have assumed this is capitalised into the cost of plant, and does not appear as an operation expense on the income statement

6.7 Estimating cost functions

In this Section, we try to estimate cost functions for subscribers, local calls and long-distance calls. In all cases, the calculations are very approximate.

The bulk of the ¢126 M plant costs reported on the balance sheet is accounted for by the Stage I investment plan, involving expenditure of ¢114 M, as set out in Table 6.7.

Our first job is to distinguish between those costs which are associated with subscribers being in the system (customer or line costs) and those costs which are associated with the making of telephone calls (call costs). Table 6.7 suggests that the bulk of local system costs (ducts, local network and installation of telephones) are of the former type, and transmission costs are of the latter type. We assume that switching equipment (including ARM) is predominantly determined by number of calls, although for local exchanges with low calling rate this would not be entirely correct. We arbitrarily assume that half the building costs are determined by number of calls and half by number of sub-scribers. Since revenue from public telephones is not included, we delete cost of installing public telephones. This gives us total capital costs of customer (line) equipment ¢49·9 M and of call equipment ¢64·1 M.

It will be convenient to reduce these capital costs to annual, and hence monthly, equivalents. The annualisation factor should take into account cost of capital, rate of depreciation, and obsolescence, mainte-nance expenses, profit etc. Rough calculations suggested that an annualisation factor of 28·85% would be appropriate.

Table 6.8 shows the result of these calculations for customer costs. It appears that average cost per line varies very considerably according to location: ¢25 in the metropolitan area, ¢43 in the provincial capitals

Table 6.8 *Calculation of average customer costs*

	Metropolitan area	Provincial Capitals	Other	Total
(i) Local exchange system (Table 6.7), ¢M	53·50	21·20	14·50	89·20
(ii) Adjustments (see text), ¢M	−24·95	−9·50	−4·85	−39·30
(iii) Adjusted capital cost (i)-(ii), ¢M	28·55	11·70	9·65	49·90
(iv) Number of lines	27 548	6480	1678	35 706
(v) Number of exchanges	5	6	9	20
(vi) Average number of lines per exchange (iv) ÷ (v)	5510	1080	186	1785
(vii) Average capital cost per line (iii) ÷ (iv), ¢	1942	3272	8641	2498
(viii) Average monthly cost per line (vii) x 0·2885 ÷ 12, ¢	24·93	43·47	138·06	33·61
(ix) Average monthly cost per exchange (iii) x 0·2885 ÷ (12 x (v)), ¢	137 333	46 944	25 741	60 000

and ₡138 elsewhere. These cost differences do not necessarily come about because the cost function itself varies by location, but more likely because different points are observed on the same function. In the smaller exchanges there are fewer subscribers over which to spread the overheads.

To calculate this cost function, it would be desirable to have data for each exchange. In the absence of such data, a rough guess can be made by using data for the three types of exchanges. If we plot monthly cost per exchange against average number of lines per exchange, the three points lie almost on a straight line. Drawing a line through the points corresponding to the metropolitan area and to the overall average yields the customer cost function:

$$\text{Monthly cost per exchange} = ₡22\,945 + ₡20\cdot76 \text{ per line} \qquad (6.4)$$

In other words, there is an overhead cost of nearly ₡23 000 to be spread over the number of lines, plus a marginal cost of nearly ₡21 per line. Thus marginal cost (assumed the same for all exchanges) is significantly lower at ₡21 than average cost, which ranges from ₡25 to ₡138, depending on the 'fill' of the exchange.

We proceed to the calculation of cost functions for calls. Relevant data from Table 6.7 are presented in Table 6.9 (local transmission costs have been arbitrarily allocated to the metropolitan area). The heading

Table 6.9   *Calculation of average call costs*

	Metropolitan area	Provincial Capital	Other	Total Local	LD	Total
½-share buildings, ₡M	4·45	3·10	1·35	8·90		8·90
Switching equipment, ₡M	20·20	6·40	3·50	30·10	2·60	32·70
Transmission, ₡M	0 90			0·90	21·60	22·50
(i) Total capital costs, ₡M	25·55	9·50	4·85	39·90	24·20	64·10
(ii) Annual equivalent (i) x 0·2885 ÷ 12, ₡M	0·61	0·23	4·85	30·96	0·58	
(iii) Average number of peak-rate calls per month, M	7·47	1·25	0·28	8·99	1·73	
(iv) Average cost per peak-rate call (ii) ÷ (iii), ₡	0·08	0·18	0·42	0·11	0·34	

LD refers to expenses incurred only for long-distance calls, the heading 'local' to expenses incurred for *all* calls. Since a long-distance call needs to pass first through the local network, the cost of a long-distance call will be obtained by adding a local component to the LD component.

Making certain assumptions about origins and destinations of calls and average durations of calls, we may deduce from information on pulses the average number of local and long-distance calls in each type of exchange. Since there is a distinct peak in the load curve, capacity has to be installed primarily to meet peak demand, and marginal cost of offpeak calls is virtually zero. It is more appropriate to calculate average cost of peak-rate calls, as in Table 6.9. The overall average cost per peak-rate local call is ¢0·11, varying from ¢0·058 in the metropolitan area up to ¢0·42 in the rural areas. The average cost of the LD component is ¢0·34. Adding this to the local components gives an average cost of a long-distance call of ¢0·45, again varying from ¢0·42 to ¢0·76 according to location.

Finally, we may repeat the same procedure as for customer costs. If we plot total costs of peak calls against average number of such calls made, for each of the three types of exchanges, these three points lie roughly on a straight line. Calculating the equation of this line yields the following cost function for local calls:

$$\text{Monthly call cost per exchange} = ¢15\,653 + ¢0·0718 \text{ per peak local call} \tag{6.5}$$

Overhead cost is thus nearly ¢16 000 and marginal cost is about ¢0·07 per peak-rate local call, which once again is substantially below average cost outside of the metropolitan area.

We do not have a breakdown of long-distance transmission costs with which to perform a similar analysis of marginal costs. One piece of information we do have is a breakdown of costs in the microwave link between San Pedro and Puntarenas:

	¢k
Site, roads, buildings	217
Power	120
Antennas and towers	138
Radio	438
Multiplex	326
	1284

Of these items, all except the multiplexing are relatively independent of the volume of traffic. Multiplex costs are incurred as additional channels are installed to meet increasing traffic requirements. At present there are 48 channels; in principle, 700 channels are possible without substantially increasing the other costs. Marginal costs are thus of the order of one-quarter of average costs (i.e. 326/128). The occupancy of this link no doubt differs from other microwave links, and they in turn differ from VHF/UHF and cable with respect to costs. To the extent that our single calculation is representative, it suggests that the marginal costs of long-distance calls are considerably below average costs in the present state of the network.

We have thus succeeded in making rough estimates of the average and marginal costs of subscribers and local and long-distance calls, differentiated by location of exchange. This information is of use for planning purposes to estimate the costs of different kinds of expansion and for purposes of setting prices. We shall later be interested to compare the costs just calculated with the prices actually charged, in order to assess the nature and extent of cross-subsidisation in the Costa Rican network.

Chapter 7

Investment appraisal

7.1 Introduction

In chapter 5 we discussed the notion of a cost function, which involved production of specified patterns of output at minimum cost. In the present chapter, we shall investigate, in a little more detail, the kinds of difficulties which arise in doing this cost minimisation. More importantly, we shall outline various ways of handling the problems which arise when comparing investments in durable equipment which yield benefits over a substantial period of time (Merrett and Sykes, 1963, Alchian and Allen, 1974, chap. 11).

Investment appraisal covers a wide spectrum of situations. At one extreme are the (comparatively) minor decisions—when to replace a worn-out piece of equipment? When to install a new exchange? What capacity microwave system to build between two locations? etc. At the other extreme are the major strategic issues—should submarine cable or satellite be used as the major vehicle for international capacity? Should obsolescent Strowger be replaced by crossbar or by electronic changes and at what rate? etc. Given the enormous magnitude of investment in telecommunications in most countries, as indicated in chapter 1, it is clearly important that investment appraisals be carried out efficiently.

Any investment appraisal involves, either explicitly or implicitly, a choice between mutually exclusive alternatives. The question may be whether or not to make a certain investment, when to install a piece of equipment, or which of several available techniques should be used. Whatever the precise situation, with each of the perceived alternatives will be associated a set of expected consequences, perhaps lasting some way into the future and by no means certain. The optimal choice will depend upon a number of factors, most importantly (i) the objectives of the decision maker, (ii) his attitude to risk and (iii) his attitude to time. We shall briefly discuss the first two factors, but the chapter is primarily devoted to the third. We return again to the first two factors in the last two chapters.

7.2   Objectives and risk

Suppose a telephone organisation is considering the replacement of Strowger by crossbar or electronic switching equipment. There are a variety of technical considerations which will limit its choice, and these need not concern us here; the point is that a choice remains. The organisation that wishes to maximise profit will very likely take a different decision from that of an organisation wishing to minimise fluctuations in employment within the supply industry and different again from that of an organisation wishing to introduce new technology as fast as possible. None of these decisions is necessarily wrong; the only criterion, from the organisation's point of view, is whether or not the decision best attains the objectives of that organisation. Thus, a given investment may be efficient for one organisation but not for another.

As a second example, consider an investment in providing a fleet of company vehicles instead of hiring vehicles from an outside supplier or using public transport. A small business, wishing to maximise profit or minimise cost, might calculate that outside supply is cheaper (using techniques to be discussed later in this chapter). A large regulated company might be compelled by the regulatory commission to hand over money saved in this way to its customers, in the form of lower prices or even a rebate, whereas an investment in a domestic transport fleet would count as 'capital' and enter the 'rate base' on which subsequent allowable profits are calculated (see chapter 13). The executives of a nationalised industry, who themselves could not benefit from any cost savings, might choose the domestic fleet for the added convenience in scheduling. Finally, a government department might prefer to buy outside in order to be seen to support public transportation.

None of these investment decisions is necessarily inefficient, if judged by the standards of the decision makers themselves. Of course, investments may be inefficient if judged by other standards, according to which certain consequences may have been given 'too much' weight, others 'too little' weight and still others may have been ignored altogether.

No one can foresee the future with certainty. Any major investment involves a degree of risk; it may be a great success or it may run into numerous unforeseen problems. Is the risk worth running, or is there a safer alternative which would be preferable? Many telephone administrations have had to choose whether to replace aging step-by-step systems by electronic or crossbar systems. Electronic systems, if successful, offer exciting new facilities, but they are relatively untried, whereas crossbar is relatively limited in scope but is well-known and efficient, and preserves a greater degree of flexibility to meet an uncertain future. The British Post Office chose to go straight for electronic switching, but the gamble did not pay off, at least in full. The Post Office had to backtrack somewhat to install some modified crossbar equipment. As it turned out, then, a better decision could probably have been made, but this was not known at the time of decision. The choice was a reflection of the British Post Office's attitude to risk; another administration might have taken a different attitude. The economist cannot say that these attitudes themselves are right or wrong. He can, perhaps, relate the attitudes taken to the type of organisation; in particular, whether it is private or public. These matters will concern us in the last two chapters.

In the present chapter we shall restrict ourselves to the investment decisions of organisations which have 'single-valued' expectations about the future. That is, they make predictions which they assume will be correct, they are subjectively certain. Various mathematical techniques have been developed for improving choice under uncertainty, notably *decision trees*, which often prove useful as a way of setting out the main alternatives; *utility functions*, which embody the degree of *risk preference* of the decision maker; and *stochastic programming* models, which enable multiple *contingent plans* to be developed. These techniques are beyond the scope of the present volume, and in any case they have not, to our knowledge, been greatly used by telecommunications entities.

The organisation may find it worth while to devote resources to obtaining information about likely changes in its environment, for example, by carrying out market research or 'industrial espionage'. It may try to prevent disadvantageous changes from taking place, or at least protect itself from being taken by surprise, by securing the commitment of customers or suppliers to a common plan, or by preventing the entry of new competitors. We shall touch on these matters again later in the book.

In the present chapter, we shall also restrict ourselves to the investment decisions of organisations which desire to maximise net revenue. Net revenue is the money proceeds accruing from the sale of goods or services, less the money outlays or expenses which were incurred in providing those goods or services. We shall assume these organisations are not concerned with nonpecuniary (i.e. nonmonetary) satisfactions such as leisure, prestige or popularity, nor do they take into account the implications of their actions for the level of employment, the balance of payments, the level of inflation etc. In a later chapter, we shall examine the decisions of organisations which choose, or are assigned, other objectives or which operate under other constraints.

An investment is a decision to use funds in one way rather than another. There are always alternative uses for funds, hence as emphasised earlier, there is always some *opportunity cost* to any investment. With our assumption about the motives of the organisation, the opportunity cost of any investment is the revenue that could have been obtained by using those funds in the best alternative way. Initially, we shall assume that the organisation has a plentiful supply of funds and that the best alternative to investing funds is to leave them in the bank. The cost of an investment is then measured by the money outlay itself (plus, as we shall see, any loss of bank interest where appropriate). At the end of the chapter, we shall explore situations where investments may only be made by raising new capital or by withdrawing funds from other profitable uses within the organisation.

7.3   The principle of time discounting

A typical investment involves a large money outlay followed by a stream of revenues (money inflows) over the next few years. How are the inflows to be compared with the outlay? The first inclination might be simply to add up all the inflows and make the investment if this sum exceeds the outlay. The limitation of this procedure is that it ignores the *timing* of the inflows.

Suppose an investment of £100 today would yield £50 next year, £60 the year after and nothing in the following years. Total revenues at £110 certainly exceed the initial outlay of £100. But suppose the firm could alternatively deposit the £100 in a bank at 10% interest rate. By next year the £100 would have grown to £110, so £50 could be withdrawn, and the remaining £60 would grow to £66 by the end of the second year, when a further £60 could be withdrawn, leaving a balance of £6. Thus, depositing the money in the bank could yield exactly the same pattern of revenues over time, and something more besides. The proposed investment is, therefore, a less profitable way of using funds.

Notice that which of the two alternatives is more attractive depends crucially on the interest rate. If the interest rate offered by the bank were only 5%, the investment would be more profitable to the bank deposit. In fact, at a bank interest rate of 6·39% the two alternatives are equivalent; at higher rates, the bank deposit is more profitable, whereas at lower rates the investment is more profitable.

It would be convenient if we could reduce £50 next year plus £60 the year after to a single amount directly comparable to the initial outlay of £100. Assuming that the interest rate is 10%, it may be calculated that a sum of £45·45 this year will grow to £50 next year (since 45·45 x 1·1 = 50); similarly a sum of £49·71 this year will grow to £60 after two years (since 49·71 x 1·1 x 1·1 = 60). Thus, at a 10% rate of interest, the sum of the two future revenues is equivalent to £45·45 + £49·71 = £95·16 today. This procedure is called *time discounting*. The rate of interest used is called the *discount rate*. The propositions involving these concepts are known as *discounted cash flow* (DCF) techniques.

Formally, if the rate of interest is 100r%, then £x today will be worth £$x(1+r)^t$ in t years' time. Equivalently, £y in t years' time is worth £$y/(1+r)^t$ today. Thus the *present value* of a stream of inflows $y_1, y_2, \ldots, y_n$ over n years, beginning in one year's time is

$$\frac{y_1}{1+r} + \frac{y_2}{(1+r)^2} + \ldots + \frac{y_n}{(1+r)^n} = \sum_{t=1}^{t=n} \frac{y_t}{(1+r)^t} \qquad (7.1)$$

This discounted sum of revenues less the initial capital cost is known as the *net present value* (NPV) of the investment. The NPV is, in effect, the profit on the investment, Thus an investment is more profitable than leaving the money in a bank at an interest rate of r% per year if and only if the NPV is positive at a discount rate of r%. Note that as regards timing of inflows and outflows of money, it is common practice to assume that capital charges are incurred before plant is brought into service, and annual costs are incurred at the end of the first and subsequent years.

If the sum of money to be received each year is constant at £y (what is called an *annuity*), expr. 7.1 simplifies to

$$y\left[\left(\frac{1}{1+r}\right) + \left(\frac{1}{1+r}\right)^2 + \ldots + \left(\frac{1}{1+r}\right)^n \right] = yS_n \qquad (7.2)$$

where S_n denotes the sum of the first n terms of the infinite series

$$S = \left(\frac{1}{1+r}\right) + \left(\frac{1}{1+r}\right)^2 + \dots \tag{7.3}$$

Multiplying both sides of eqn. 7.3 by $(1+r)$ yields

$$(1+r)\,S = 1 + \left(\frac{1}{1+r}\right) + \left(\frac{1}{1+r}\right)^2 + \dots$$

$$= 1 + S$$

hence

$$S = \frac{1}{r} \tag{7.4}$$

Thus at a discount rate of $100r\%$, the present value of a perpetual annuity of £y per year, beginning at the end of year one, is y/r. For example, a perpetual annuity of £1000 per year is worth 10 000 at a 10% discount rate, twice that at a 5% discount rate and only £5000 at a 20% discount rate.

To obtain an expression for the finite annuity S_n, we may split the infinite series S into two parts:

$$S = S_n + \left(\frac{1}{1+r}\right)^{n+1} + \left(\frac{1}{1+r}\right)^{n+2} + \dots$$

$$= S_n + \left(\frac{1}{1+r}\right)^{n} \left[\left(\frac{1}{1+r}\right) + \left(\frac{1}{1+r}\right)^2 + \dots \right]$$

$$= S_n + \left(\frac{1}{1+r}\right)^{n} S$$

hence

$$S_n = \left[1 - \left(\frac{1}{1+r}\right)^{n} \right] S$$

and, using eqn. 7.4

$$S_n = \frac{1 - \left(\dfrac{1}{1+r}\right)^{n}}{r} = \frac{(1+r)^{n} - 1}{r(1+r)^n} \tag{7.5}$$

Tables are available (e.g. Morgan, 1976) giving the present values of single payments and annuities, for wide ranges of discount rates r and dates of payment n.

7.4   Two examples

Suppose a telephone company can manufacture and install a flexible telephone extension cord for $20. The company proposes to rent out the extension cord at $3 per annum, payable in arrears, over its estimated 8-year life. Is this a profitable venture? Although the simple (undiscounted) sum of revenues is $24, the total discounted revenue at 10% is

$$\$3\left[\left(\frac{1}{1\cdot1}\right) + \left(\frac{1}{1\cdot1}\right)^2 + \ldots + \left(\frac{1}{1\cdot1}\right)^8\right] = \$3 \times 5\cdot335 = \$16$$

(where $S_8 = 5\cdot335$). This is not a profitable proposition since the NPV $= 16{-}20 = -\$4$. However, it would be profitable at a rental of about $20/5\cdot335 = \$3\cdot75$, or even at the original rental if the life of the cord exceeded 12 years (since 12 is the lowest value of n for which S_n exceeds 20/3).

Often the revenues accruing to two mutually exclusive projects are the same, notably where two different methods of provision of the same service are under consideration. In this case the projects may be compared on the basis of the minimum present value of cost. It must be borne in mind, however, that this criterion does not ensure that even the cheapest alternative considered is more profitable than doing nothing, or putting the money on deposit at the bank.

Consider an example due to Scherer (1970, p. 534). Suppose two coaxial cables are needed, one immediately and one in ten years' time. The two can be buried simultaneously for an outlay of $25 000 per mile. Alternatively, one cable can be buried today at an outlay of $20 000 per mile and the trench can be reopened ten years later to bury the second cable at an additional outlay of $10 000 per mile. Which alternative is cheaper? Evidently the answer depends on the discount rate, since the present value of cost under the second alternative is $20 000 + $10 000/(1+r)^{10}$. The following table compares the two alternatives at different discount rates:

Present value of cost (per mile)

Discount rate	Bury simultaneously	Bury sequentially
%	$	$
0	25 000	30 000
4	25 000	26 756
7	25 000	25 084
10	25 000	23 855

At discount rates of 7% or less, burying both cables simultaneously is the cheaper alternative; at discount rates over 7%, the opposite is the case.

This brings out another general point about discounting. If the interest rate is high because funds are scarce, expenditures and receipts in the near future are weighted very heavily compared with future outlays and inflows. It becomes more attractive to postpone capital investment and to choose projects with quicker payoffs. By contrast, as interest rates fall, one can afford to indulge in longer-term projects. It is, therefore, important to know what is the appropriate discount rate to use, and to revise one's plans as interest rates change.

7.5   Alternative methods of evaluation

A concept which is often encountered is the *internal rate of return* (IRR) of a project. This is defined as the rate of interest (or discount rate) which would be required to yield a NPV of zero. To decide whether the investment is worth while, this IRR is compared with the interest rate at which money can be raised or loaned. In most cases, the two procedures will give the same results. Many businessmen prefer IRR because it avoids the need to specify an exact interest rate; one merely needs to know whether alternative opportunities will yield more or less than the IRR. However, there are certain difficulties associated with the IRR. For instance, it may not be unique if the cash flow contains alternating positive and negative terms. In such circumstances the NPV can exhibit a 'switching' phenomenon—at low discount rates alternative A is cheaper, at higher rates alternative B is cheaper, and at yet higher rates alternative A is once again cheaper.

Another method is to reduce capital costs to equivalent annual costs. Thus at an interest rate of 5% a capital cost of £100 is equivalent to an annual cost of £5 in perpetuity, or to an annual cost of £8 (of which £5 is interest and £3 capital repayment) over a 20-year life. This element may be added to other annual costs such as maintenance, and alternative schemes thereby compared. Once again, this method should give the same answer as the other two. It is useful where there is only a single initial capital charge, but it becomes complicated where there are many such changes, the timing of which differ between the alternatives being compared.

During the 1960s the use of IRR and NPV methods probably increased. In 1967 the British government specified that '. . . . the [nationalised] industries are free to use whichever method suits them best, but the government will expect projects which are submitted to it for approval to be expressed in present values by the use of a test-rate of discount' (Cmnd 3437, para. 8).

The British Post Office apparently prepares data on annual costs for all its standard items, and finds it convenient to work with the 'present value of annual cost' (Morgan, 1976). This appears to amount to NPV and it is not clear what advantage the more circuitous procedure offers.

Regardless of the exact method of appraisal used, the following fundamental point emerges. Expenditure and savings today are more important than expenditures and savings tomorrow. The further into the future we look, the less weight should be attached to future outcomes. The reason is not that the future is unimportant, nor that it is uncertain. Rather, the reason is that money invested today can generate additional wealth by tomorrow; withdrawing that money today sacrifices a greater future wealth. Putting the future into its proper perspective will lead to more, not less, wealth in the future.

7.6 Inflation

Inflation is a general increase in the level of prices. Until the 1970s inflation in most developed countries proceeded at a rate of $1-2\%$ per annum, which could safely be ignored. Nowadays, rates of inflation up to 15% per annum are not uncommon in Europe and in 1975 inflation reached 26% in Britain. Of course, in Latin America and many developing countries inflation rates of around 100% per annum are frequently encountered.

There is no denying that inflation is a nuisance in making investment appraisals. It is also dangerous because it can give a falsely optimistic view of any investment. Take the previous example of renting a telephone extension cord, and suppose that the company expected prices *including rentals*, to rise by 10% per annum. The present value of these rentals, with an interest rate of 10%, would be

$$\$3 \left[\frac{1\cdot1}{1\cdot1} + \left(\frac{1\cdot1}{1\cdot1}\right)^2 + \ldots + \left(\frac{1\cdot1}{1\cdot1}\right)^{10} \right] = \$30$$

The rate of inflation just offsets the discount rate and the proposition now appears attractive.

But of course the higher rental in future years will not buy more then; in *real* terms $3·30 next year is worth the same as $3 without inflation. One obvious approach is to work in real terms by deflating future anticipated revenues by the rate of inflation. This is what is meant by calculations *in real terms*, or in *constant pounds or dollars*. This procedure is reasonably standard nowadays.

If calculations are carried out in real terms, the rate of interest should also be the real rate, but this will differ from the current money rate. If lenders require a real return of 5%, and if they expect prices to increase by 10% they will demand a *nominal* (money) interest rate of 15% to compensate them. Thus, if the rate at which one borrows correctly reflects the rate of inflation, the nominal interest rate should be deflated to a *real* level by subtracting the expected rate of inflation.

There are, however, further difficulties because expectations about the future rate of inflation generally differ. During 1975, inflation in Britain was over 25%, but nominal interest rates were about 15%. Anyone lending money would have been better off than anyone keeping it in his pocket, but in real terms he would still have been about 10% worse off at the end of the year than at the beginning. The real rate of interest was about minus 10%. People may have been willing to lend at 15% because they did not expect inflation to be as high as 25%, or they may not have had access to more profitable opportunities for using money. Whatever the reason, any stock which merely maintained its value in real terms (i.e. whose price increased at the rate of inflation) would have been a profitable investment at interest rates then available.

The test discount rate prescribed by the British government in 1967 for appraisal of nationalised industry investment was set in real terms. Originally, it was set at 8% and it was subsequently increased to 10%. These figures were intended to be comparable with rates of return earned in private industry. In the last few years *nominal* rates of return in industry have been higher, but it is believed that *real* rates of return have been much lower. In 1978 the test discount rate was supplanted by a required rate of return of 5% in real terms before tax, intended to bring the public and private rates more closely into line (Cmnd 7131). The objection to reducing the rate earlier was that, if effective, it would have stimulated investment by nationalised industries at a time when a cutback in expenditure was sought. Such matters are discussed later in the book.

The question of inflation is clearly a difficult one. Morgan (1976, pp. 4,103) in fact warns against taking it into account:

It has been shown that the rates of return determined in DCF studies involving income and expenditure are very sensitive to differential rates of inflation. This sensitivity of the results to even small changes in the fore-cast (assumptions) of inflation rates can easily be used to manipulate the economic study to prove in or justify a preferred scheme. If use is made of forecast rates of inflation it is essential that sensitivity analysis techniques are applied rigorously but in general cost comparisons over a long study period are likely to be less suspect if the inflation/deflation situation is ignored.

In a period of rising prices there is always the temptation to adjust the study to take account of inflation but this should be resisted. Adjustments for inflation always favours the schemes with the high initial capital cost. An alternative which can be justified without any allowance for inflation will only appear more attractive when the inflation allowance has been made, whereas if the adoption of a scheme is dependent principally on the allowance for inflation its justification is suspect.

This advice appears to conflict with our own, but if we understand the author correctly, this is not in fact the case. He is in effect suggesting that appraisals should be made in real terms as we have proposed and his warning about the biasing of calculations should be heeded. For further discussion about techniques of investment appraisal in infla-tionary circumstances, the reader is referred to the specialised financial literature, e.g. Merrett and Sykes. (1963). There is also currently under discussion in Britain the question of 'inflation accounting', which has been adopted for use by British nationalised industries.

7.7   The cost of capital

We have hitherto assumed that the company has sufficient funds on hand to finance any contemplated investment, and that the best alternative to investing the funds in any project is to leave them on deposit in the bank.. Consequently, the (opportunity) cost of an invest-ment project is the revenue which the necessary money outlay would generate if left in the bank, i.e.. the revenue which would be lost by investing in the project.

Suppose, however, that the firm's funds were fully committed to other projects. Then the question arises—which is the cheapest way to obtain the necessary funds—to borrow from a bank or some other external source, or to cut back on some current project within the company? Suppose that $10 000 is required and that the bank would loan it at a rate of 10%, necessitating the payment of $1000 interest next year. Suppose also that the installation of $10 000 worth of new trunks somewhere in the system could be postponed, which would lose a certain amount of revenue from lost calls. Ignoring for the moment the possible loss of goodwill, the internal source of funds is to be preferred if and only if the loss of revenue is less than $1000. This, in turn, would be worth doing if and only if the proposed new project was likely to generate more revenue than the revenue sacrificed by postponing the installation of new trunks.

The cheapest rate at which funds can be acquired is called the *cost of capital*. Expressed as a percentage or proportion, it is this rate which should be used as the discount rate in net-present-value calculations.

The cost of capital to the firm will depend on where that capital has to be raised, or on the mixture of sources which are used. The calculation of this cost of capital is a topic on which there has been a great deal of argument in the United States, since the regulated companies there are allowed sufficient revenues to cover this, but (in principle) no more. We shall not go further into this topic, except to stress the important point that the cost of any investment depends on the price which has to be paid currently (or will have to be paid) for the new capital necessary to carry out the investment, i.e. on the alternative uses which are currently foregone. It does *not* depend on the price which was paid at some time in the past. That price is now a sunk cost. For further discussion, see the papers in Trebing and Howard (1969) and the references therein.

If the organisation has no access to outside funds, or only limited access, it faces the problem of allocating what limited funds it has on hand among a set of available projects. This is known as the *capital budgeting* problem. In recent years, mathematical techniques such as linear programming have been applied to this problem (e.g. Weingartner, 1966, Wilkes, 1977 and the references therein). They have proved useful in practice and have yielded new insights into the nature of the problem. For example, there emerges from a multiperiod linear program a set of 'shadow prices', one for each constraint. The shadow price on the constraint pertaining to the limited availability of funds in year t indicates the marginal opportunity cost of capital in that year, i.e. the return which could be earned with a little extra capital, or which would be sacrificed if a little capital were withdrawn. This is useful practical knowledge, since it indicates when additional borrowing should be sought, and what price is worth paying; it also enables new projects to be costed out and compared with existing projects. It is immediately apparent that the appropriate discount rate is not necessarily constant over time. There may be an acute shortage of funds in one year compared to the excellent opportunities available, whereas in another year there may be surplus funds available.

In the present chapter, we have for simplicity assumed a constant discount rate, but it is now clear that, in principle, a different discount rate for each year might be appropriate. It will usually not be worth making elaborate calculations with different calculations. However, failure to acknowledge this possibility sometimes leads to confusion. For example, it was mentioned earlier that from 1967 to 1978 the British government required nationalised industries to evaluate major investment projects according to an 8% (later 10%) test discount rate. This rate was supposed to reflect the return earned in private industry. The purpose of the exercise was to ensure an appropriate division of resources between the public and private sector. In this case, the government should have approved all projects which showed a positive NPV at the test discount rate. In fact, it did not, because it was willing to make available only limited funds. In some years the industries were told that even projects which passed the test could not be accepted; in other years, the government wished to stimulate the economy by 'bringing forward' investment projects which were originally planned for later years (and were therefore unprofitable, in that accepting them now meant sacrificing a higher return later). The test discount rate and the NPV procedure became largely irrelevant. The Post Office, in particular, suffered from the resulting wild swings in government policy and investment levels.

The root of the difficulty was that the industries were being asked to use a discount rate which did not in fact correspond to the opportunity cost of capital available to them. It was often suggested at the time that a correct solution could be obtained by raising or lowering the test discount rate to reflect the opportunity cost of capital. This was held to be unacceptable, partly because the resulting rates would have been embarrassingly high or low, and partly because they would have misrepresented the cost of capital in later, more 'normal' years. The present discussion suggests that an alternative solution would have been preferable; instead of maintaining a single discount rate, why not raise or lower the discount rate for this (exceptional) year only, while maintaining a 'normal' rate for later years?

Chapter 8

Pricing policy

8.1 Introduction

In the last few chapters we have paid scant attention to the question of price. Much of the discussion of investment appraisal concerned the cheapest way to provide a specified service, implicitly assuming that the demand for that service was somehow determined exogenously. However, we have already established in chapter 3 that the demand for any service will respond to the price of that service. The question arises, therefore, as to how telecommunications organisations should set their prices. The purpose of this chapter is to explore that question.

Conventional accounts by economists of the determination of price and outputs usually begin with the case of *perfect competition* where there are sufficiently many producers of the same product that each firm must take the market price as given, and merely adjust its output. Attention then shifts to industries where, for one reason or another, there is only a single producer, a *monopolist*, who can choose his price in the light of the demand curve facing him (Lipsey, 1971, Ferguson and Gould, 1975).

In many ways, it is simpler and more natural to begin with the monopolist, who is the first to discover a market. Later, he is subject to competition from newer entrants into the industry. One may then study the effects of competition on prices and outputs, and understand the original producer's attempts to fend off or outdistance his competitors. Emphasis on the case of monopoly is particularly appropriate for telecommunications, where competition within a country is the exception rather than the rule.

8.2   Total, average and marginal revenue functions

Recall from chapter 3 that a demand curve $x = x\,(p)$ expresses the relationship between the price p of any commodity and the number of units x of that commodity which would be demanded at that price. We expect that it will be downward sloping, suggesting that more units will be demanded at a lower price. To continue the example of the previous chapter, suppose the telephone company estimates that 8000 subscribers would be willing to buy a telephone extension cord at a price of \$13, and 10000 at a price of \$8. At an even higher price of \$20 perhaps only 5200 would be willing to buy. It is convenient to represent demand as a continuous function of price, and the above three points all fall on the straight-line demand curve

$$x = 13\,200 - 400p \tag{8.1}$$

The total revenue which is generated at any price varies with the price. At a price of \$8, demand is 10000 cords, hence total revenue is \$80000; at a price of \$13, demand is 8000, hence total revenue is \$104000; at a price of \$20, demand is 5200, hence total revenue is again \$104000. The general relationship is

$$R(p) = px(p) = p(13\,200 - 400p) = 13\,200p - 400p^2 \tag{8.2}$$

Maximum revenue is attained where

$$\frac{dR(p)}{dp} = 13\,200 - 800p = 0 \tag{8.3}$$

hence at $p = 13\,200/800 = \$16 \cdot 50$. At this price 6600 cords will be demanded and total revenue will be \$108900.

To compare revenue curves with cost curves, it is convenient to express them both as a function of the same variable, namely output. Accordingly, let $p(x)$ be the *inverse demand curve*, defined as the highest price at which x units could be sold. Here

$$p(x) = 33 - x/400 \tag{8.4}$$

The *total revenue function* (now as a function of output) is

$$R(x) = xp(x) = 33x - x^2/400 \tag{8.5}$$

as shown in Fig. 8.2. The rate at which total revenue increases as output is increased is known as *marginal revenue*, hence

$$MR(x) = \frac{dR(x)}{dx} = p(x) + \frac{xdp(x)}{dx} = 33 - x/200 \qquad (8.6)$$

The marginal revenue curve is the *slope* of the total revenue curve. the latter attains its maximum where the former is zero. The marginal revenue function is shown in Fig. 8.1. Note that the *average revenue function $AR(x)$* is simply the (inverse) demand function, since

$$AR(x) = \frac{TR(x)}{x} = p(x) \qquad (8.7)$$

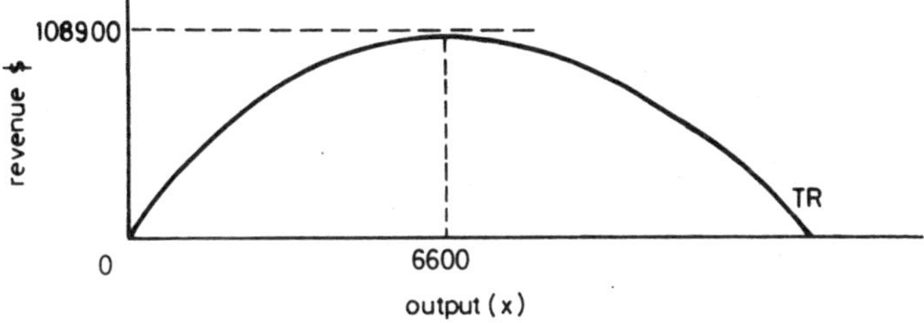

Fig. 8.1   *Total-revenue curve*

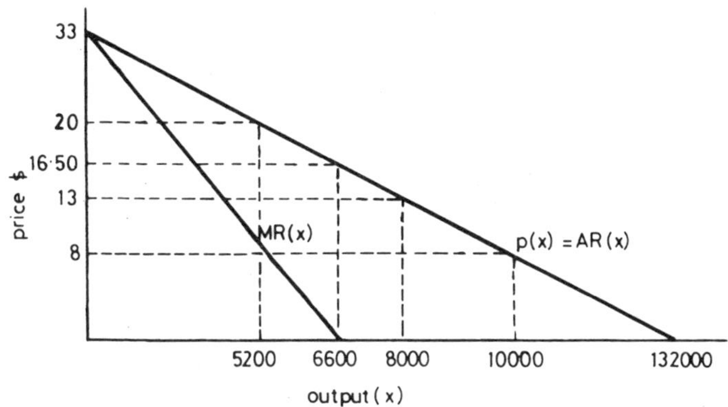

Fig. 8.2   *Demand curve and marginal-revenue curve*

With straight-line demand functions, the marginal revenue function is also a straight line with twice the slope of the demand function, starting from the same point on the vertical (price) axis.

The concept of price elasticity of demand can usefully be related to revenue. Recall from Section 3.5 that price elasticity is defined by

$$e = \frac{dx}{dp}\frac{p}{x} \tag{8.8}$$

Marginal revenue is thus given by

$$MR(x) = p + x\frac{dp}{dx} = p\left(1 + \frac{1}{e}\right) \tag{8.9}$$

When total revenue is at a maximum we have

$$MR(x) = 0 = p\left(1 + \frac{1}{e}\right), \text{ hence } e = -1 \tag{8.10}$$

If price elasticity is less than unity in absolute value, i.e. if demand is inelastic, total revenue will be increased by raising price; if demand is elastic, it will be increased by lowering price. Where there is unit elasticity, total revenue is independent of price.

Note, however, that elasticity will generally vary with the level of price (i.e. it will vary along the demand curve). With the previous example of the linear demand curve $x = a - bp = 13\,200 - 400p$, we have

$$e = \frac{-bp}{a-bp} = \frac{-400p}{13\,200-400p} \tag{8.11}$$

Elasticity takes the values $e = -0 \cdot 44$ at $p = \$10$, $e = -1 \cdot 0$ at $p = \$16 \cdot 50$, $e = -1 \cdot 54$ at $p = \$20$ and $e = -10$ at $p = \$30$. In fact, absolute elasticity increases from 0 to infinity as price increases from zero to the price of $\$33$ at which demand falls to zero.

8.3   Pricing for maximum net revenue

If the firm wishes to maximise net revenue, the optimal output will be where total revenue less total cost reaches a maximum. Let $C(x)$ be the total cost of producing x units; the net revenue function, denoted $\pi(x)$, is

$$\pi(x) = R(x) - C(x) \tag{8.12}$$

which reaches a maximum where

$$\frac{d\pi}{dx} = \frac{dR}{dx} - \frac{d(C)}{dx} = 0 \tag{8.13}$$

or where marginal revenue $MR(x)$ equals the marginal cost $MC(x)$.

These curves are illustrated in Fig. 8.3.

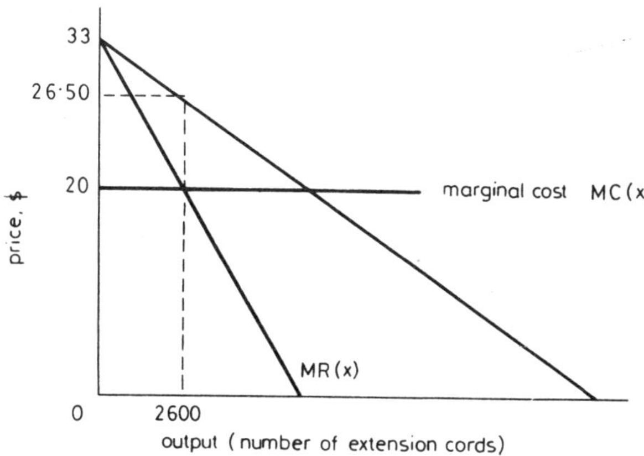

Fig. 8.3   *Setting price and output to maximise net revenue*

To continue the former example, if the telephone extension cords can be produced for $20 each, $C(x) = 20x$ and $MC(x) = 20$. Setting marginal revenue equal to marginal cost yields $33 - x/200 = 20$, hence optimal output is $x = 2600$ cords. This number of extension cords can be sold for a price of $p(2600) = 33 - 2600/400 = \$26 \cdot 50$ each. At this price, elasticity is $e = -400(26 \cdot 5)/[(13\,200 - 400(26 \cdot 5)] = -4 \cdot 08$.

By expressing the optimality condition of eqn. 8.13 in terms of elasticity of demand, it is straightforward to show that the optimal price satisfies

$$p\left(1 + \frac{1}{e}\right) = MC \text{ or } \frac{p}{c} = \frac{e}{1+e} \tag{8.14}$$

In the present example, $p/c = 26 \cdot 0/20 = 1 \cdot 325$ and $e/(1+e) = 4 \cdot 08/ 3 \cdot 08 = 1 \cdot 325$.

To take another example, Naoe (1974) has calculated that the price elasticity of demand for international calls from Japan is $-2 \cdot 28$ (Table 3.2). Suppose this elasticity is thought to be approximately constant in the region of present prices. It is easily calculated from the formula above that maximum net revenue on such calls will be attained when $p/c = -2 \cdot 28/(1+2 \cdot 28) = 1 \cdot 8$, that is, when prices are set at about 80% above the marginal cost per call.

Observe that almost all the price elasticities of demand for telephone calls and connections reported in Tables 3.1 and 3.2 are less than unity. This suggests that telephone prices are generally not set to maximise net revenue, since in such cases net revenues could be increased by raising prices until demand became elastic (see also chapter 10 for a further examination of Illinois Bell's pricing policy).

8.4   Price discrimination

Price *discrimination* is the sale of the same product at two or more different prices (Alchian and Allen, 1973). It is profitable for the firm to discriminate if it can separate its customers into two or more classes whose elasticities of demand differ appreciably. For example, suppose it were thought that residential subscribers in Japan are much more sensitive to the price of international calls than are business customers; assume the demand elasticities are $-4 \cdot 0$ and $-1 \cdot 5$, respectively. If different prices were to be charged for international calls, the optimal price markups for the two classes of customers would be

$$\text{residential } \frac{p}{MC} = \frac{-4}{1-4} = 1 \cdot 3, \text{ business } \frac{p}{MC} = \frac{-1 \cdot 5}{1-1 \cdot 5} = 3$$

To maximise net revenue, prices to residentail subscribers should be 30% above marginal cost, while prices for business subscribers should be three times marginal cost. At these prices, marginal revenue will be the same for both classes of customer.

A substantial difficulty with implementing such a policy is that businesses would attempt to make international calls from residential phones, or would even attempt to purchase a residential line. Successful price discriminataion requires that buyers can be prevented from shifting to the cheaper price. Presumably, this is why telephone companies almost never differentiate between customers with respect to price of calls. An apparent counter example from the USA is the sale of telephone channels on a bulk basis under the TELPAK tariff. It has been calculated that in 1967 a buyer taking only 12 channels paid over six times per channel what a buyer taking 240 channels paid. Nevertheless even the smallest bulk buyer paid only one-sixth the effective price charged to customers paying by the call (Littlechild and Rousseau, 1973, p.40). Small industrial and domestic customers would not have the traffic to justify the dedicated lines involved in TELPAK, and FCC regulations at the time prohibited the parcelling together of small users to take advantage of the bulk rate.

In many countries there are price differentials between business and residential subscribers for installation fees and subscription rates (rentals). Of the 55 countries whose telephone tariffs are reported by

Roos *et al.* (1976), 10 differentiate their installation fee by type of subscriber, eight by size of local network and nine by some other method; for monthly rentals, the corresponding figures are 20, 17 and 18 countries. Italy, for example, has a special category for 'state, regional, provincial, municipal and governmental, authorities and schools, press agencies, newspapers and people who have journalism as their only or principal activity' (See also B.M. Mitchell, 1978*a*).

Although these differentials may reflect a difference in elasticities of demand (i.e. discrimination), it might also be argued that costs of serving the various classes of customer are different. The concept of price discrimination is frequently extended to cover such cases by defining discrimination as the sale of two or more similar goods at prices which are in different ratios to marginal cost (or, alternatively, whose difference in prices is not equal to differences in marginal cost). Whether business and residential rentals are discriminatory then turns on whether the difference in prices exceeds the difference in costs, on which we have no evidence.Thus different prices need not necessarily reflect discrimination.

Conversely, equality of prices does not demonstrate the absence of discrimination. In the United States, and indeed in most countries, long-distance calls over heavy traffic routes are much more profitable (because of economies of scale) than long-distance calls over light-traffic routes. Since the same price per mile is charged, people making calls over the heavy-traffic routes are, in effect, being discriminated against.

The most common form of price differentiation for calls is by time of day: 37 out of the 55 countries in the survey by Roos *et al.* (1976) practised this policy. Whether this involves price discrimination is not immediately apparent, because the costs of providing capacity are not separable, but instead are joint over the different periods of the day. The principles of peak-load pricing are discussed in more detail in chapter 10.

It is quite common practice to discriminate over time, by introducing new products at high prices then gradually reducing these prices after the most eager buyers have made their purchases. This is done with books (where the difference in price between hardback and paperback versions presumably exceeds the difference in binding costs), with films at first-run then local cinemas, and with General Motors automobile accessories gradually moving down from the Cadillac range to the Pontiac range. The pricing of optional equipment for telephone subscribers, such as colour telephones in the UK, follows a similar pattern.

Two final points should be made in this Section. The major criticism of monopoly, from an economist's point of view, is that output is restricted to increase the profit of the producer. Resources are not utilised in the way in which consumers would prefer, and in the way in which competition would ensure. For if competition were feasible, prices could not be set above costs; profits would be competed away. This criticism of monopoly is, strictly speaking, only valid when the monopolist is constrained to charge a single price. If he is able to discriminate amongst all his customers, under most circumstances he will be able to increase his output. If he can discriminate perfectly, his output will be identical to that of a competitive industry. Of course, his profit will be much higher as a result of discrimination, which may not be desirable, but the point to be made here is that the level of production will not be artificially restricted.

It may also be the case that production would not be possible at all unless price discrimination could be practiced. Suppose the annual cost of operating a small telephone exchange were £270 000 plus £100 per subscriber. Suppose further that the annual demand curves for connection by business x_1 and residential x_2 subscribers were respectively

$$x_1 = 11250 - 12 \cdot 5p_1 \text{ and } x_2 = 13333 - 33 \cdot 3p_2$$

where p_1 and p_2 refer to the annual rentals. It may be verified that maximum net revenue would be obtained by charging annual rental of £500 to business subscribers and £100 to residential subscribers, at which rates 500 members of each class would subscribe. The elasticities of demand at these prices are $-1 \cdot 25$ and $-1 \cdot 67$, respectively. Net revenue (profit) is just £5000 per annum. The exchange would still break even if rentals were reduced slightly, but there is no uniform rental which could be charged to all subscribers that would still allow the exchange to break even. Without price discrimination (or a subsidy), this telephone exchange could not survive.

8.5   Two-part tariffs

Like the products of most utilities, telephone services are commonly sold under a 2-part tariff. As discussed in chapter 2, the subscriber pays a *fixed charge* (or rental) for the use of a telephone plus a *variable charge* (or price) per call made. In fact, the tariff typically has many parts, since a variety of additional services and pieces of equipment may be purchased or rented. In addition, an initial connetion fee is usually levied, and possibly a disconnection or moving fee (see, for example, the Costa Rican tariff summarised in Section 6.2).

The intriguing aspect of a 2-part tariff is that the two parts are necessarily closely intertwined. One cannot set price per call to maximise net revenue on calls without at the same time affecting demand for connections, and conversely. Considering its practical importance, there has been relatively little discussion of optimal 2-part tariffs in the economic literature (see, however, Pigou, 1920, Lewis, 1949, chap. 2, Hazlewood, 1950, Oi, 1971, Littlechild, 1970*a,b*, 1975*a,b*, Pyatt, 1972, Squire, 1973, Leland and Meyer, 1976). We give here an intuitive idea of the results for a firm maximising net revenue.

The optimal rule, as before, is to set the rental fee for connections such that the marginal revenue from an additional subscriber equals the marginal cost of connecting him. Marginal revenue is now the net addition to income from rentals (taking into account that the rental charge to all subscribers must be slightly reduced to attract one additional subscriber) plus the net revenue on calls which the additional subscriber makes. Formally, we have

$$f\left(1 + \frac{1}{e_f}\right) + (p-c)\,x = g$$

where f is the fixed charge or rental per unit time, e_f is the elasticity of demand for connections with respect to this fee (at a given set of call prices), g is the marginal cost per unit time necessary to connect a subscriber (i.e. cost of telephone instrument, cable pair to the local exchange and terminal equipment there), $p-c$ is the net revenue (price less marginal cost) per telephone call and x is the number of calls made by the marginal subscriber per unit time. The left-hand side of this equation is marginal revenue, the right-hand side is marginal cost, both with respect to number of subscribers.

At the same time, the price of a call should be set so that the marginal revenue from an additional call equals the marginal cost of providing that call. Account must also be taken of the effect on rentals. If call prices are reduced, the telephone system becomes more attractive and a given subscriber would be willing to pay a higher price to join. Consequently, increasing the number of calls by reducing their price generates some additional revenue in the form of more rentals (or, alternatively, in more subscribers at the same rentals). Formally,

$$p\left(1 + \frac{1}{e_p}\right) + N\,\frac{\partial f}{\partial x} = c$$

where e_p is the elasticity of demand for calls with respect to the price of a call, N is the number of existing subscribers and $\partial f/\partial x$ is the rate of change in rental with respect to the number of calls (a change made possible by the lower price of calls). Again, the left-hand side is marinal revenue of a call, the right-hand side is marginal cost thereof.

Whether such formulas can be used numerically in a real telephone system remains to be seen. Perhaps their most useful contribution is to formalise the insight that demands for connections and calls are interdependent. The marginal revenue of supplying either is somewhat higher than might appear at first sight; consequently, the telephone system should be expanded, by charging lower prices and rentals, rather more than the examination of calls or subscribers, taken separately, would suggest.

These mathematical models also reveal interesting new possibilities (Oi 1967, Littlechild, 1975*b*). Telephone administrations may find it profitable to set subscription charges *below* marginal subscriber costs (f less than g, in our notation), to make profits on the calls which additional customer make. Alternatively, it may even be profitable to make a 'loss' on calls (p less than c) to charge a higher subscription rate. As we shall see later, there is some evidence that the price of local exchange service (i.e. subscription plus local calls) is set below cost, being compensated by profits on long-distance calls. Various explanations are put forward for this policy; one possibility is simply that it is the most profitable policy.

As pointed out earlier, it is sometimes argued that a new subscriber joining the telephone network confers a benefit on other subscribers, in that their choice of people to call is now increased. The telephone company may find it profitable to take this into account. The implications of such externalities are discussed in chapter 12.

8.6   Competition and regulation

We have so far examined the optimal pricing policy for a firm in a monopoly position wishing to maximise net revenue. There are basically three sources of such monopoly power:

(*a*)   The firm may be the only one to recognise the profitability of the product; others could compete, but, as yet, are not inclined to do so. This is the usual state of affairs when new products and techniques are being developed, not least in telecommunications. Apparently the American firm Xerox was unsuccessfully offered to over 40 British firms before Rank decided to take the gamble (which subsequently turned out to be extremely profitable and attracted numerous competitors).

(*b*)   The firm may be the sole owner of resources necessary to produce the product. Insofar as technical knowledge is such a resource, this is true of aspects of electronic switching and digital transmission.

(*c*)   The firm may be protected from competition by law or government regulation. The patent system is a notable example, so too is the statutory monopoly of posts and telecommunications enjoyed by the British Post Office. In the USA, the FCC has, until recently, effectively prevented new entrants from challenging the existing telecommunications firms.

We shall leave these issues of government policy for a later chapter. Here, we wish to trace the effects of competition which is allowed to proceed (Alchian and Allen, 1973). As competitors learn of the succes of any firm they will attempt to emulate this success and capture the existing profits by offering a similar product at a lower price or a better product at a similar price. Competition of this kind is evident between Bell and the independent US telephone companies, and even between European administrations to a limited extent (although in the latter case the motivation is primarily prestige rather than revenue). Active price cutting, improvements in quality, and the advertising of these new developments, are the hallmarks of the competitive process by which each firm tries to outdo its rivals.

Traditional economics has developed the concept of *perfect competition* to describe the (surely hypothetical) culmination of this competitive process. Assume there are many producers of a homogeneous (i.e. identical) product, all with perfect knowledge about market conditions of demand and supply. Only one price can rule in such a market, since customers would not pay a higher price than could be obtained from a rival firm. Moreover, the output of each firm is supposed to be small relative to the size of the market as a whole, so that each firm can get all the business it wants at the going market price. Each firm produces whatever quantity of output is most profitable at the going price; this is the output where marginal cost equals price, which in this type of market is equal to marginal revenue, since what the firm produces does not affect price. Competition forces this market price down to the level of minimum average cost, which is the same for all producers, since they are all assumed to have access to the same cost functions (Lipsey, 1971, Ferguson and Gould, 1975).

The situation thus obtained, called *competitive equilibrium*, has several desirable properties, including (i) industry output is distributed among the firms in such a way as to minimise total costs, (ii) price is equal to marginal cost of production (which is the same for every firm), (iii) total output of the industry is at the level which maximises value of output to consumers less cost and (iv) the expectations of the firms about prices and demand are confirmed, so that plans do not need to be revised. As remarked earlier, there are no monopolistic restrictions of output, and price discrimination cannot be sustained.

This notion of competitive equilibrium lies at the heart of welfare economics, to which we now turn.

Chapter 9

Welfare economics and marginal-cost pricing

9.1 Introduction

Economics, or at least that branch of it called *positive* economics, is concerned to explain how the economy works. But for many years economists have wished to give advice to governments as to what *ought* to be done. They have wished to identify what kind of legal, political and institutional framework should be provided for the economy; when the government should intervene to supplement, guide or replace the working of the private economy; what principles should be followed in making such interventions, at what levels to set tax rates, by what rules to operate nationalised industries, whether or not to make specific investments etc. During this century there has gradually been developed a substantial body of *normative* economic theory dealing with these issues, which goes by the name of *welfare economics* (Graaff, 1957, Millward, 1971). An important aspect of welfare economics is *marginal-cost pricing*. In this chapter we shall describe the basic approach of welfare economics, indicate how it relates to official government policy, and spell out some of the implications for pricing and investment policy in telecommunications.

9.2 The optimality of perfect competition

Under certain assumptions, it may be shown that an economy which is perfectly competitive will produce efficiently those goods and services which consumers demand (given their incomes). Moreover, with the resulting allocation of resources, it will be impossible to make anyone better off without making at least one person worse off. Such a state of affairs is said to be *Pareto efficient* or *Pareto optimal*, after the eminent

Italian economist Vilfredo Pareto (1848–1923). The argument is briefly as follows:

There are several requisites for *perfect competition*, notably
(*a*)   A large number of producers of each commodity, no one of which is large enough to affect the market as a whole by his own actions, so that each producer must take the market price as given.
(*b*)   Perfect knowledge by producers about relevant techniques of production and prices of inputs, and perfect knowledge by consumers about prices of consumer goods, so that inefficient decisions are not made in error.
(*c*)   Absence of externalities in production and consumption, which would cause the decisions of producers and consumers, which are based on their own private interests, to diverge from the interests of society as a whole.

If these conditions hold, the production and price of each commodity are determined by the intersection of the demand and supply curves for that commodity. Production of any given amount is undertaken by those producers who have the lowest costs, and production and consumption are carried forward to the point where the benefit of an extra unit is equal to its price, which in turn is equal to its marginal cost. In this way, the optimal amount of each commodity is produced in the most efficient way. In the resulting equilibrium situation, it will not be possible for any producer to reallocate his resources into some other more profitable line of production, nor will any consumer be able to do better by changing his consumption plan. Society is said to be on its *production possibility frontier*. The only way of making one consumer better off is to transfer more income to him from some other consumer, or to impose other restrictions on the economy which will likewise harm somebody else. This is what is meant by *Pareto optimality*.

9.3   The role of government

In such a perfectly competitive economy, central planning could not do any better for everybody simultaneously. In fact, given that the intention is to carry out the wishes of consumers, resources are allocated precisely where a omniscient central planner would allocate them. This has not been achieved by central direction, but by many individual firms responding to profit opportunities revealed by the price mechanism. Perfect competition assures that all these opportunities are noted and exploited, and the profit is competed down to a 'normal' rate reflecting the cost of capital.

What, then, is the role of the government? Broadly speaking, there are two justifications for intervention. The first is that the outcome of a perfectly competitive economy, although efficient, may not be socially desirable for one reason or another. Redistribution of income is perhaps called for. The second justification is that the conditions for perfect competition may not be achieved in a particular market. This is a case of *market failure*, and government intervention is required to correct this failure. In this view, one role of the economist is to identify such areas of market failure and to prescribe appropriate corrective action (Meade, 1975). Consider in turn the three prerequisites for perfect competition:

(a)   Number of producers

Suppose there are only a few producers in a particular market. This may be because economies of scale are considerable relative to the size of the market. If production is to take place on an efficient scale, there is room for only a few producers. Presumably, this is the case in the steel and car industries of all countries. Each producer is then conscious that other producers will be aware of his own price, and may well retaliate to any move he makes. This situation is called *oligopoly*. There is an incentive for the firms in each industry to collude in order to raise prices, and with only a few firms it may not be too difficult to make and police a suitable arrangement. There is evidence that such *cartels* have operated from time to time in the electrical manufacturing industries of the USA and Great Britain. Collusion has also taken place in the British telecommunications supply industry (the *Ring*), but it has usually been with the approval of the Post Office; we return to this issue in the last chapter.

Whether or not explicit collusion actually takes place, it is felt that competition is likely to be less strong if there are fewer producers. As a result, prices will be above marginal cost, 'supernormal' profits will be earned, demand will be diverted (by higher prices) to other industries, so that output will be below the optimal level in this industry and above the optimal level elsewhere. Resources will have been misallocated.

The appropriate remedy depends on the severity of the problem. It may be sufficient to prohibit mergers in the industry, or to allow mergers only subject to guarantees of suitable behaviour, or it may be necessary to investigate the industry, perhaps to condemn certain practices or perhaps even to break it up. Alternatively, the industry may be subject to regulation or nationalisation. This is typically the solution adopted when the industry is thought to be a *natural monopoly*, i.e. when only one single firm can operate in a market at an efficient scale. Telecommunications is generally held to be a natural monopoly; in the USA the industry is regulated and in the UK it is nationalised.

The principles dictated by welfare economics for the control or operation of such industries may be summarised quite briefly. Essentially, *the price of each product or service should be set equal to its marginal cost*, and output should be expanded to meet resulting demand at those prices. There should be no *cross-subsidisation*, whereby one service is charged at a price above marginal cost to finance the supply of another service at a price below marginal cost. An investment in new capacity should be undertaken if and only if the present value of the additional net income stream which it generates is positive, where income and expenditure are discounted at a rate equal to the *social opportunity cost of capital*, that is, the return which that capital could generate in its best alternative use elsewhere in the economy.

(b)   Perfect knowledge

It is well known that some firms are more efficient, use more up-to-date techniques, and are more ready to innovate, than are other firms. Presumably, the former have knowledge about some techniques that the other firms lack. One reason given for the creation of the government-financed National Enterprise Board in Britain is to make these backward firms more efficient.

A second possibility of market failure is that firms will not, in general, know the future. They cannot be sure what products consumers will want and in what quantities, what proportion of this demand will be met by other firms, whether suppliers will have the necessary capacity etc. Consequently, one justification for a national plan, or for a series of planning agreements, is to spread and co-ordinate knowledge of plans between firms. If bottlenecks and overcapacity can be avoided, resources can be used more efficiently.

(c)   Externalities

If firms find a cheaper method of production which incidentally pollutes the river or the atmosphere, they will not hesitate to use it unless restrained in some way. Fish may be driven out of the river and consumers may also suffer through having to wash clothes or repaint houses more frequently. These losses may more than outweigh the gains enjoyed by the firm. From society's point of view, the new process is inefficient. It is, therefore, suggested that the government intervene, either by prohibiting or limiting pollution, or (preferably) by bringing the costs of pollution to bear upon the polluter by means of a tax. When the externalities are thus *internalised*, the polluter himself will make the correct decision, from society's point of view, as to whether to go ahead with this new technique. There are evidently certain difficulties of income distribution here, for although, in principle, the gains will more than outweigh the losses, those who lose are not necessarily compensated by those who gain.

The use of *cost-benefit analysis* is frequently advocated and adopted to handle important investment decisions involving externalities (Layard, 1972, Irvin, 1978). The idea is to identify, evaluate and aggregate all the costs and benefits, to whomever they accrue, which follow from any decision under consideration. Examples include the report of the Roskill Committee (1971) on the Third London Airport and various water supply studies in the USA (e.g. Hirshleifer *et al.*, 1960). In chapter 12 we shall explore more thoroughly the existence and implications of externalities in telecommunications, notably with respect to congestion or grade of service, and with respect to size of the network. We shall also examine the need to modify financial calculations in a developing country where resource prices may be very different from social opportunity costs.

9.4 Marginal-cost pricing as official Government policy in Britain

Since the 1930s economists in many countries, especially Britain, France and the USA, have been engaged in heated debate about the appropriate pricing and investment policies for nationalised and regulated industries. By the 1960s it seems that a significant number of politicians and civil servants in Britain had been convinced of the merits of welfare economics. In 1968 the House of Commons Select Committee on the Control of Nationalised Industries presented the case for marginal-cost pricing as follows:

> There is also the requirement that prices should reflect costs so that a transfer of resources from one activity to another can raise consumer welfare. If consumers are willing to pay more for some extra output than it costs to produce, welfare will *prima facie* rise if that expansion of output and sales takes place. In the reverse case, if consumers are unwilling to pay a price that covers the costs of producing marginal output, welfare will rise if a contraction of output and sales takes place. This is the rationale behind a policy of marginal-cost pricing. [Select Committee, 1968, p.47].

The unexpectedly heavy investment requirements of the nationalised industries during the early 1960s focused attention on the inadequacy of existing guidelines for control. In the 1967 White Paper entitled 'Nationalised industries: a review of economic and financial objectivities' (Cmnd. 3437, para. 4), the Chancellor of the Exchequer observed that

> . . . calls upon scarce resources are now very heavy; and the need to measure these calls, to assess priorities, and to allocate resources upon economically and socially rational basis has become even more important.

The White Paper went on to specify four aspects of policy:

(a)   *Investment*

> Investment projects must normally show a satisfactory return in commercial terms unless they are justifiable on wider criteria involving an assessment of the social costs and benefits involved, or are provided to meet a statutory obligation (para. 7).

Discounted cash flow techniques were recommended for all important projects, and

> . . . the government will expect projects which are submitted to it for approval to be expressed in present values by the use of a test rate of discount. [para. 8]

It was decided that 8% was a reasonable discount rate to use, which was thought to be broadly consistent with the average return (in real terms) looked for on low-risk projects in the private sector. This was intended to achieve an efficient allocation of resources between the public and private sectors. The implications of this policy for investment appraisal have already been discussed in chapter 7.

(b)   *Prices*

The first principle was that the revenues of nationalised industries should normally cover their accounting costs in full. But this was not sufficient:

> . . . pricing policies should be devised with reference to the costs of particular goods and services provided. Unless this is done, there is a risk of undesirable cross-subsidisation and consequent misallocation of resources. The aim of pricing policy should be that the consumer should pay the true costs of providing the goods and services he consumes, in every case where these can be sensibly identified. [para. 18]

There were exceptions to this general rule–for commercial reasons, impracticability, statutory requirements, wider economic or social considerations.

> But–these cases apart–to cross-subsidise loss-making services amounts to taxing remunerative services provided by the same organisation and is as objectionable as subsidising from general taxation services which have no social justification. [para. 18]

There followed some discussion of pricing policy in particular situations. Where there was persistent spare capacity, price should be reduced, if necessary to the level of escapable cost, otherwise investment might wrongly be stimulated in competing industries. Where there was a regular cycle of demand, offpeak price should be reduced and peak price increased. Two-part tariffs and differential pricing systems could be used where common costs were significant.

It was further stated that 'prices need to be reasonably related to costs at the margin' (para. 21), and there was some discussion of the conditions under which short-run or long-run marginal cost was more relevant.

(*c*)   *Costs*

It was insufficient merely to ensure that prices properly reflect costs; 'management's task is to seize every opportunity of reducing costs by employing improved methods or techniques', especially by making labour savings.

(*d*)   *Financial objectives*

> Clear financial objectives will continue to be necessary so that the industries know what is expected from them by the government. Thus they serve both as an incentive to management and as one of the standards by which success or failure over a period of years may be judged [para. 33]

These targets were generally set as a percentage return on net assets.

The 1967 White Paper reported that henceforth all price increases proposed by nationalised industries were to be referred to the National Board for Prices and Incomes (NBPI). Evidently, the NBPI had no doubt about the appropriate criterion for pricing policy in the Post Office; for in its 1968 report on *Post Office Charges* it firmly declared

> The Post Office should consistently follow a long-run marginal-cost approach for the establishment of future pricing policy, and for this pur-

pose should set up a team of engineers, statisticians, economists, operational researchers and cost accountants to undertake the necessary work in order to identify the structure of long-run marginal costs. [p. 61]

In practice, as we shall see in chapter 14, there was little enthusiasm towards marginal-cost pricing. Nevertheless, the marginal-cost pricing philosophy was further endorsed by the Report of the Post Office Review Committee chaired by Mr. Carter (1977). Further, a new White Paper issued in 1978 (Cmmd 7131 p. 23) explained 'how the Government intends to reintroduce and reinforce the approach to investment appraisal, pricing policy and financial targets which was set out in the 1967 White Paper'.

9.5   The implications for telecommunications pricing

In a paper originally published in 1950, Hazlewood had already outlined the kind of pricing policy for telephone systems that would result. We may summarise his proposals as follows.

(*a*)   A subscription or rental per subscriber per period of time (e.g. per month) should equal the marginal cost of maintaining a subscriber in the system. These customer costs consist of the capital cost, converted to a monthly basis, of the subscriber's handset, drop wire, cable pair to the local exchange, and that part of the switching equipment therein which is solely associated with that customer. To these elements should be added the cost of meter reading and rendering regular accounts. Customer costs would typically differ for each subscriber, notably according to distance from the exchange, but it will not be practicable to differentiate the charge quite so finely.

(*b*)   There may be certain indivisibilities in customer costs, for example, the cost of much exchange equipment and outside plant, such as power and ventilating equipment, building and land, which are not immediately attributable to any particular subscriber, but rather to all subscribers in a particular area. Dividing these common costs equally among subscribers might exclude some potential subscribers willing to pay the marginal cost incurred in connecting them. The balance of common costs should, therefore, be recouped according to 'what the traffic will bear', i.e. price discrimination. This might well involve charging a higher rental to businesses than to residential customers.

(*c*)   Traffic costs (of switching and transmission capacity) are primarily related to *peak* traffic, and should be allocated to peak calls only, according to route. Offpeak calls should be charged only customer cost (wear and tear on the equipment, power, metering and billing costs etc.).

(*d*)   Flat-rate pricing, where the charge per call is zero, is criticised on several grounds: it encourages a wasteful use of the telephone, insofar as calls are made which would not have been if the cost of them had been charged; small users subsidise large users; and some potential small users are kept off the system by the high charge. Analogous arguments are made against schemes where the monthly subscription is zero and all costs are recouped from call charges.

(*e*)   A part of the cost of a call is incurred only in setting up the call and is unaffected by duration. However, during the peak period, capacity is utilised which prevents other calls being made. This suggests that prices of offpeak calls be independent of duration, but not so for peak calls (although the price for successive minutes may be less than for the first minute).

(*f*)   The value of a telephone service resides essentially in the number of connections it offers. It might be advantageous to charge less than full customer cost to some (potential) subscribers whose connection to the system is thought to yield particular external benefit to other subscribers.

9.6   Graphical and mathematical representations

It is usual to analyse the questions of marginal-cost pricing by using simple graphs and mathematical formulations of cost and demand functions. In the present Section we shall examine the pricing of a single service (subscriber rentals). The next Section discusses the modifications induced by financial constraints. The next chapter explores the intricacies of peak-load pricing of telephone calls, where joint costs are involved. Externalities are dealt with in chapter 12.

Let us divide the costs of providing local exchange service into those that are incurred in building and equipping the exchange, which are assumed to depend on the design capacity of the exchange, and those that are incurred in operating the exchange, which are assumed to depend on the number of subscribers enlisted. We shall refer to these as *capacity costs* and *operating costs,* respectively. We shall not be concerned in this Section with those costs which depend on the volume of traffic, nor with the effect of subscription charge on volume of traffic.

We wish to determine first how a telephone administration following the principles of welfare economics should set its subscription charges and number of subscribers, given that it has inherited an exchange of a certain fixed capacity. We then determine the optimal exchange capacity to design in the first place. Assume for simplicity that exchange operating costs are linear in number of subscribers x, so that $\pounds c$ is the constant marginal operating cost per subscriber per unit time up to fixed capacity y_0 subscribers. This is shown graphically in Fig. 9.1. Suppose that demand curve D_1 holds. The height of the demand curve at any point indicates the value (per unit time) which the most eager additional subscriber would place on being in the network.

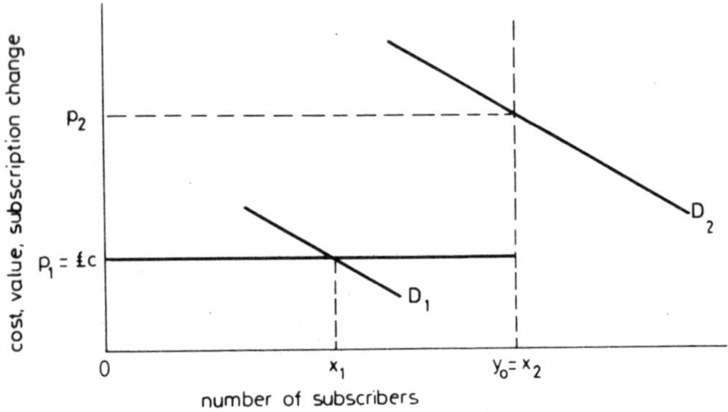

Fig. 9.1   *Determining optimal subscription charges and number of subscribers with fixed exchange capacity*

Optimal subscription charge and number of subscribers are now determined as follows. Increase the number of subscribers as long as the marginal evaluation of membership exceeds the marginal operating cost per subscriber, up to the level x_1, where the marginal value of a subscriber equals the marginal cost. At this number of subscribers the net value of the system (i.e. total value to connected subscribers less total costs of operation) are at a maximum. To generate this number of subscribers, the subscription charge p_1 should be set equal to the marginal operating cost per subscriber ($\pounds c$).

Now consider the case where demand the curve D_2 holds. To maximise the net value of the system, the number of subscribers should be increased to the level of capacity ($x_2 = y_0$). However, if the subscription charge were set equal to the marginal operating cost per subscriber in an exchange with spare capacity, there would be excess demand and waiting lists for telephones. This has the further disadvantage that the available lines would not necessarily be allocated to those subscribers placing the greatest value on them. The solution is to set the subscription charge at the level p_2 which just clears the market, i.e. where the number of customers wishing to subscribe just equals the capacity available.

Does this market-clearing price bear any relation to marginal cost (except insofar as one may interpret the marginal cost curve as vertical at capacity output)? The following interpretation is often given. If an exchange is operating at full capacity, the only way of adding a new subscriber is to remove one of the existing subscribers. The existing subscriber who values membership the least currently values his membership at p_2 as given by the height of the demand curve at full capacity. This value has to be sacrificed if a new subscriber is admitted, and therefore constitutes the *marginal opportunity cost to 'society'* of adding a new subscriber. In this way the principle of marginal-cost pricing may be preserved, by assuming that the telephone administration acts to maximise value to 'society', where value to society is defined as the simple aggregate of money values to its members.

Thus the general principle is that the subscription charge should be set so as to make best use of the capacity available. It should be set equal to marginal operating cost per subscriber or to a market-clearing level, whichever is higher. If there are no overriding financial constraints, the sunk cost of building the exchange is irrelevant. For the purpose of setting subscription charges, the only relevant costs are those to be incurred in the future, in operating the exchange. Charges set on this basis will not necessarily equal financial outlays, nor will they necessarily maximise profit.

Suppose we now examine the problem at an earlier stage. What is the optimal capacity to design for? A similar principle applies: capacity should be extended to the point where its marginal value equals its marginal cost. With demand curve D_1 and capacity y_0, additional capacity has no value; there is surplus capacity already. With demand curve D_2, on the other hand, the marginal value of an additional unit of capacity is the price p_2 at which it could be rented out less the operating cost c. This is often called the *marginal opportunity cost of capacity*. With demand curve D_2, capacity should therefore be greater or less than y_0, according to whether $p_2 - c$ is greater or less than marginal capacity cost at y_0, i.e. whether marginal opportunity cost of existing capacity at y_0 is greater or less than marginal construction cost of new capacity.

Assume for simplicity that marginal capacity cost per subscriber is constant at £g per unit time. The exchange should be designed for x^* subscribers, where the value which the x^*th subscriber places on being in the system is exactly equal to marginal capacity cost c plus marginal operating cost g; the optimal subscription charge p^* is equal to this same sum. Formally, if $p(x)$ denotes the inverse demand curve for connections, we have

$$p* = p(x*) = c + g$$

Demand at this subscription charge just equals exchange capacity, as shown in Fig. 9.2.

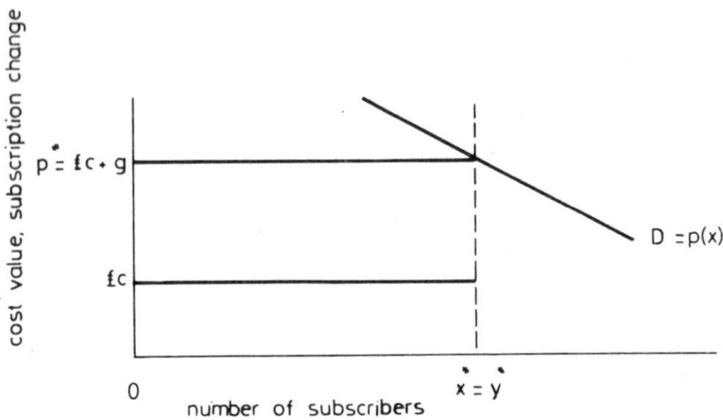

Fig. 9.2   *Determining optimal exchange capacity*

This procedure may easily be demonstrated mathematically. Recall that $p(x)$ denotes the inverse demand curve, that cx is the total operating cost of x subscribers and that gy is the total cost of an exchange designed for y subscribers. The problem is to choose design capacity and number of subscribers so as to maximise net value of the system, given by total value of membership less total costs of building and operation, subject to the constraint that number of subscribers does not exceed design capacity. Formally,

$$\underset{x,y}{\text{maximise}} \quad \int_0^x p(x')\,dx' - cx - gy \tag{9.1}$$

$$\text{subject to } x \leqslant y; x \geqslant 0, y \geqslant 0 \tag{9.2}$$

This is a simple nonlinear programming problem (see appendix to chapter 10). The solution is to design capacity $y*$ equal to expected membership $x*$ at the point where

$$p(x*) = c + g \tag{9.3}$$

If, on the other hand, an initial capacity y_0 is fixed, the problem simplifies to

$$\underset{x}{\text{Maximise}} \quad \int_0^x p(x')\,dx' - cx \qquad (9.4)$$

$$\text{subject to} \quad 0 \leqslant x \leqslant y_0 \qquad (9.5)$$

which has the solution

$$p(x^*) = c + u^* \qquad (9.6)$$

where u^* is a *dual variable* or *shadow price* or *Lagrange multiplier* corresponding to the capacity constraint. This shadow price equals the rate at which the objective function can be increased as the initial level of capacity is increased; it is equal to zero if there is spare capacity and equal to the height of the demand curve less the height of the marginal curve otherwise. In fact, it is precisely the marginal value or opportunity cost of capacity described earlier, where the level of capacity may be designed to be optimal. This marginal value u^* is precisely equal to marginal capacity cost g, as in eqn. 9.3.

In practice, there is a significant lag between the time at which a telephone exchange is designed and commissioned and the time at which it is open for business. The question is therefore often raised— should the subscription charge be set at the level envisaged when the exchange was designed (i.e. equal to marginal capacity cost plus marginal operating cost) or at the level appropriate at the time of opening (i.e. equal to marginal operating cost plus the marginal *opportunity* cost of capacity)? Now if the forecast of demand proved to be accurate, so that the exchange was built to the right level of capacity, there is no difficulty; the two rules give the same answer. However, if forecasting errors have occurred, the two rules may well give different answers, and economists differ in their prescriptions for policy. Some economists argue that, in the light of the model developed hitherto, it is correct to ignore the sunk costs and wrong forecasts, and make the best use of available capacity in the light of present demand. Other economists would argue that the basic model is oversimplified, that in practice frequent fluctuations in price to meet changing demand are inconvenient, expensive and confusing for customers, hence the charge envisaged at the design stage should be maintained despite demand fluctuations. Yet other economists advocate some form of compromise, for example, recalculating costs and prices at appropriate intervals of, say, two years (Boiteux, 1949, Turvey, 1969, Kahn, 1970, Vickrey, 1971, Mills, 1976).

9.7   Financial constraints

In both Britain and the USA, regulations provide that revenues should, above all, be sufficient to cover accounting costs and allow a specified rate of return on capital. It is not at all uncommon to find constraints of this kind on public enterprises. They serve to ensure that customers of the enterprise, taken as a whole, pay their way, and are not subsidised by customers of other products or by taxpayers. Not only does this seem equitable, it also ensures, in a rough-and-ready way, that only those services are provided which customers wish to pay for.

On the other hand, a financial constraint may well conflict with the requirements of marginal-cost pricing. If marginal cost is increasing with the number of subscribers, total revenue will exceed total cost, which raises the question of what should be done with the profit. On the other hand, if there is a substantial overhead cost, so that marginal cost is everywhere below average cost, then a loss will be made. It is conceivable, but unlikely, that the various profits and losses on the different services of some enterprise will just cancel out. If this does not happen, how should the principle of marginal-cost pricing be reconciled with that of breaking even?

One possibility is to mark up all prices by an equal amount, or by an equal percentage. Another possibility is to discriminate according to 'ability to pay' or 'what the traffic will bear', as in Hazlewood's suggestion, described in Section 9.5. The rule to be described in this Section is a mathematically rigorous way of doing just that. It has been advocated by a number of economists, but is particularly associated with the name of Frank Ramsey. A discussion of the history of the rule is given by Baumol and Bradford (1970).

Suppose it is required to set the prices of n different services so as to maximise net value of output subject to earning net revenue R. Let $p_i(x_i)$ be the inverse demand curve for service i and let c_i be its marginal cost (assumed constant for ease of exposition). The problem is to choose outputs x_i and implicitly p_i equal to $p_i(x_i)$, so as to

$$\text{maximise} \quad \sum_{i=1}^{n} \int_{0}^{x_i} p_i(x_i') \, dx_i' - c_i x_i \tag{9.7}$$

subject to the net revenue constraint

$$\sum_{i=1}^{n} x_i \, [p_i(x_i) - c_i] \geq R \tag{9.8}$$

Let the *Lagrange multiplier* (or shadow price, see appendix to chapter 10) of the revenue constraint be denoted λ. Intuitively, it measures the rate at which net value can be increased as the revenue constraint is relaxed. The optimal solution is then given by

$$(p_i - c_i) + \lambda \left[(p_i - c_i) + x_i \frac{dp_i}{dx_i} \right] = 0 \quad \text{for all } i \tag{9.9}$$

The second term in the bracket on the right may be written $p_i e_i$, where e_i is the price elasticity of demand for the ith service (which, recall, is defined by $e_i = dx_i/dp_i / p_i/x_i$). Since price elasticity is negative, it will be more convenient to use its absolute value $|e_i| = -e_i$. We can then rearrange the above expression to read

$$\frac{p_i - c_i}{p_i} = \frac{\lambda}{1 + \lambda} \frac{1}{|e_i|} \quad \text{for all } i \tag{9.10}$$

This is *Ramsey's principle of optimal pricing*; the resulting prices are called *Ramsey prices*. For all services, the profit margin should be the same fraction of the inverse of price elasticity of demand. This means that where demand is inelastic, price should be set higher than where demand is elastic, in a very precise way. If price elasticity of demand for connections by residential subscribers is twice as high (in absolute terms) as by business subscribers, rentals should be set so that the profit margin (not the rental itself) is twice as high on the business rental.

Another way of looking at the rule is that the *Ramsey numbers* a_i, defined by

$$a_i = \frac{p_i - c_i}{p_i} e_i \tag{9.11}$$

should be constant, and equal to $\lambda/(1+\lambda)$, for all services.

Willig and Bailey (1977) have made some calculations of the Ramsey numbers obtaining on the US interstate message toll telephone service in 1973. They took 21 milage bands (varying from short haul to long haul) for three times of day. Price elasticities and cost coefficients were based on preliminary experimental data, and crosselasticities of demand (e.g. between different times of day) were ignored. These data showed that demands were more elastic for longer-haul calls than for shorter ones, and more elastic for evening and night calls than for day calls.

A plot of the Ramsey numbers is shown in Fig. 9.3. If the telephone company were following the Ramsey principle of relating prices to inverse elasticities, one would expect all the Ramsey numbers to be the same. Thus, one would expect higher profit margins on short-haul and daytime calls than on long-haul and night-time calls. In fact, the Ramsey numbers increased steadily with length of haul, and day time Ramsey numbers were just under half the corresponding evening and night-time ones.

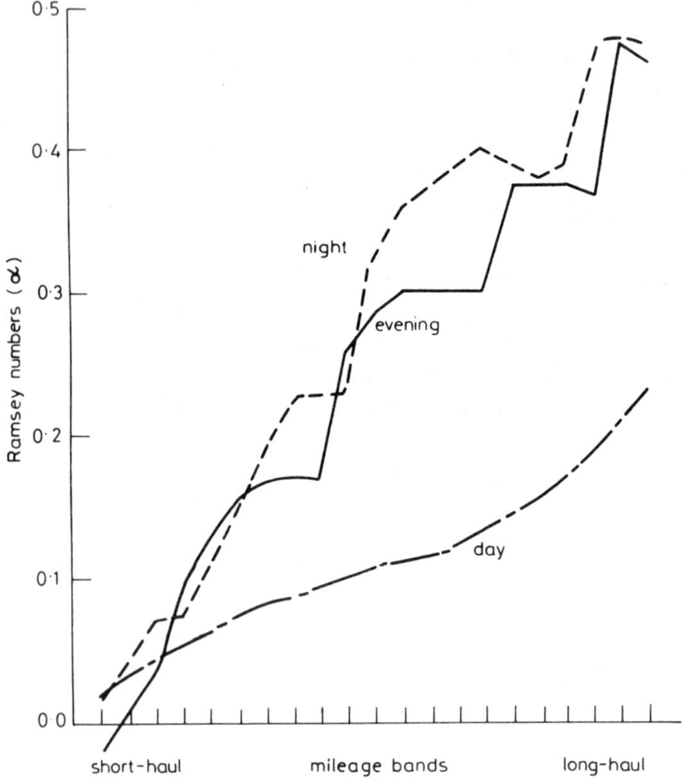

Fig. 9.3   *Ramsey numbers of 1973 DDD rate schedule (USA)*
[Source: Willig and Bailey]

If the underlying data are correct, these results indicate that the telephone company is not maximising net value of output subject to a net revenue constraint. The Ramsey number of about zero for all short-haul calls implies that in 1973 no profit at all was being made on those calls. A Ramsey number of 0·5 for night-time long-haul calls suggests a profit of 50% of price if demand has unit elasticity, 100% of price if demand has elasticity 0·5 etc.

At the same time, it should be noted that the 1973 prices are well below profit-maximising levels. It is not difficult to show that profit maximisation implies

$$\frac{p_i - c_i}{p_i} = \frac{1}{e_i} \qquad (9.12)$$

that is, a Ramsey number equal to 1. The results of Willig and Bailey suggest that profit margins rose to at most half the profit-maximising levels, and for short-haul calls were well below the proportion.

According to Ramsey's theory, the welfare of customers taken as a whole could have been increased by substantially increasing the price of short-haul calls and reducing the price of long-haul calls. The price of day-time calls should have been increased by more than night-time calls. The authors calculate that the improvement in net value of output would have been somewhere between \$69 M and \$249 M (depending on the shape of the demand curves). As a percentage of current value of output this would have been between $0 \cdot 8\%$ and $2 \cdot 9\%$. These ranges are so wide that one cannot place too much confidence in them, but the characterisation of actual pricing policy is of considerable interest. A similar calculation is described in the next chapter, which is not entirely consistent with these results.

Before leaving the subject of financial constraints, it should be noted that there is another means of raising revenues which may yield less loss in value of output. It has so far been assumed that a single price must be charged for each service. However, multipart and declining-rate tariffs are not uncommon in telecommunications. It would be possible to charge a certain rate per call for a specified number of local calls, then a lower rate thereafter. Providing a subscriber makes more than the specified number of calls, his calling rate is predominatly determined by the price of the marginal call, and the higher charge for the first few calls is a way of extracting revenue without substantially affecting his calling pattern. However, a qualification must be made here because empirical evidence from the British Post Office[*] appears to suggest that many consumers regulate their calling pattern according to the size of their total bill, rather than the price of a call at the margin.

9.8 Flat-rate versus 2-part tariffs

We mentioned earlier the argument by Hazlewood, which would probably be endorsed by all economists, that local exchange services

---

[*] Private communication

E.T.F — K

should be charged according to a 2-part tariff. It may be shown formally that, under the usual assumptions of welfare economics, subscription rate should be set equal to the marginal cost of keeping a subscriber in the system, and a charge per (local) telephone call should equal the marginal cost of such a call. If a minimum profit or financial return has to be achieved, profit margins on both rentals and calls should be increased *in the same proportion* until the desired level of profit is achieved (Littlechild, 1975*b*).

In practice, however, many telephone administrations adopt a flat-rate tariff for local exchange service, i.e. the rental charge includes an unlimited number of (free) local calls. The outstanding example is the United States. According to Roos *et al.* (1976), at least 15 other countries follow this policy. The historical justifications for flat-rate tariffs (e.g. to generate the 'calling habit') may no longer be valid. Economists have argued that flat-rate tariffs artificially stimulate the number of calls, necessitating an increase in rental which deters smaller subscribers from joining the network. There is increasing interest in the USA in the possibility of a move to a 2-part tariff, or *usage-sensitive pricing*, as it is called there (Garfunkel, 1975, Alleman, 1977, B.M. Mitchell, 1978*b*).

The chief economic argument in favour of flat-rate pricing, and against a change of policy, is the *cost of metering* local calls and billing the subscriber accordingly (Matthewson and Quirin, 1972). Metering costs are likely to vary considerably with the size of exchange, the type of equipment used and the exact requirements of metering (e.g. whether calls are to be timed as well as counted). B.M. Mitchell (1978*b*) reports estimates that the cost per line could be over $50 per annum in small mechanical exchanges with only a few hundred subscribers, whereas the New York Telephone Co. has apparently installed metering equipment in its large crossbar exchanges for $15 per line. The same company reports that converting large electronic exchanges to metered service should cost no more than $2 to $5 per line. Estimates of record keeping and billing costs vary from $0·001 to $0·003 per call, or 10 cents to 40 cents per subscriber per month for the average number of local calls.

Mitchell has attempted to calculate the effect of changing from a flat-rate to a 2-part tariff and to evaluate the social benefits of doing so. He uses scattered information on cost and demand functions such as we have reported in earlier chapters, and carries out extensive sensitivity analyses. In all cases, a shift to a 2-part tariff leads, as expected, to the attraction of new subscribers, especially from low-income groups, and to a reduction in number of calls per subscriber, especially by low-income subscribers. Under a 2-part tariff, people who make substantially more than the average number of calls will pay higher telephone bills, while low-volume callers will save money.

The net gains to society depend crucially on the costs of metering. Where these are high, they may more than offset the benefits of a 2-part tariff. In electronic exchanges, however, gains may be about 9% of current net benefits (i.e. consumer plus producer surplus). Mitchell estimates that in the residential sector alone the use of 2-part (usage-sensitive) tariffs coupled with peakload pricing of local calls could achieve welfare gains of the order of $500 M per year.

9.9   Charging for directory assistance

In 1970, the New York Public Service Commission rejected the New York Telephone's proposal for 'across-the-board' rate increases, and suggested that rates for each service be more closely aligned with costs. In 1971, the Commission asked the company to address itself to several specified aspects of service, including charging for excessive use of directory assistance (DA):

> The company has informed the Commission that 50% of calls to directory assistance are made by less than 20% of all callers and indicates a predominance of business usage of this service. The company is still studying the feasibility of imposing some sort of charge for excessive use of this service by business users, but has not as yet made any definite proposal. Directory [assistance] service charges would appear to be an area in which several desirable objectives can be attained, namely, reduction of usage, operational savings, reduced capital requirements and attraction of additional revenues. [quoted in Stannard, 1978, p. 84]

Despite considerable opposition, a charging plan of 10 cents per call to directory assistance, with certain exemptions and credit for making fewer than six calls per month, was put into effect on 1 September 1975.

This proposal provides an excellent opportunity to compare the criteria of profit and consumer surplus as a basis for decision making. The following discussion is based on the paper by Stannard (1978) and subsequent comments by Alleman (1978), although it will become apparent that the evaluation given here differs from theirs, and gives a substantially lower estimate of the net social benefits of DA charges.

In round figures, the relevant magnitudes appear to be as follows. Directory assistance calls initially cost an average of 16·5 cents per call to provide. The imposition of a 10 cents charge per call increased revenue by $10 M per annum and reduced costs by $12 M per annum. For our purposes, we may assume that there were initially 173 M calls eligible for the charge, of which 73 M were eliminated leaving 100 M paying the charge (in fact, there were some 500 M calls to directory assistance, but many of these were not eligible for the charge). This situation may be represented by straight-line demand and cost curves, as in Fig. 9.4. We also ignore here the increase over time in number of directory-assistance calls made.

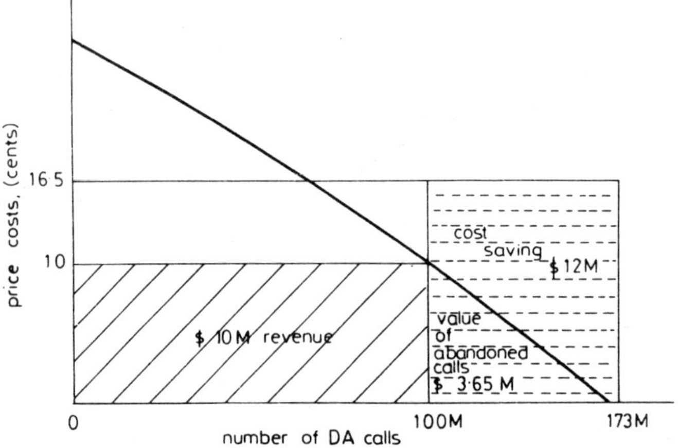

Fig. 9.4   *Analysis of directory assistance charging scheme: effect on DA calls*

From the company's point of view, the total net benefit of the charging scheme is apparently $22 M per annum, comprising the revenue of $10 M plus the cost savings of $12 M.

The average customer views the scheme rather differently. For those 100 M directory assistance calls which he continues to make, he now has to pay $10 M. He also foregoes 73 M calls because they are not worth 10 cents each; assuming they are worth an average of 5 cents each means that he has lost benefits worth 73 M x $0·05 = $3·65 M per annum. Customers in total are thus $13·65 M worse off as a result of the charging scheme.

Now the $10 M revenue is merely a *transfer* of income from customers to the telephone company. The net saving to society as a whole (consumers plus telephone company shareholders) is the cost saving of $12 M less the value of the foregone calls $3·65 M, yielding a net saving of $8·35 M per annum.

The analysis is not yet complete, however. Assuming that the telephone company is effectively regulated, it will have to reduce its revenue requirement elsewhere by precisely the $22 M gained from the charging scheme. Suppose, for purposes of illustration, that the telephone company currently charges a price for long-distance calls just sufficiently above their marginal cost to finance the total cost of directory assistance calls, namely 173 M x $0·165 = $28·5 M. Assume, for simplicity, that each toll call costs $1 to provide, but that a price of $1·10 is charged, and that 285 M toll calls are made per year, thereby exactly financing the cost of directory assistance.

With the scheme for charging directory assistance calls adopted, the price of toll calls must be reduced so as to generate a lower 'profit' on toll calls of only 100 M x $ (0·165 - 0·10) = $6·5 M. The appropriate level of price will depend on the elasticity of demand for toll calls. Assume demand is unit elastic (so that total revenue is independent of price—see Section 8.2) It may be calculated that a new toll price of $1·021 is required. At this price, demand for toll calls is 307 M and a 'profit' of (nearly) $6·5 M is generated. This situation is shown in Fig. 9.5

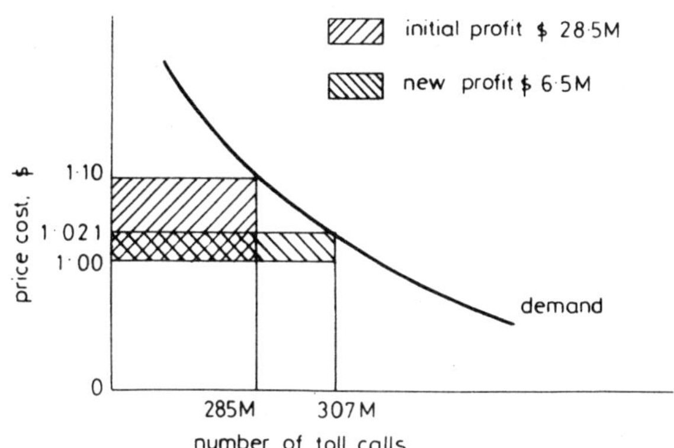

Fig. 9.5   *Analysis of directory assistance charging scheme: effect on toll calls*

Compared with the previous situation, the new situation represents a transfer of income from the telephone company to customers making the original toll calls, in amount $(1·10 - 1·021) x 285 M = $22·5 M. The company makes up an additional $0·5 M profit on the 22 M toll calls, and these additional calls generate a consumer surplus of approximately ½$(1·10 - 1·021) x 22 M = $0·87 M. The total gain on toll calls is thus $0·5 M + 0·87 M = $1·37 M.

The net effect of introducing the directory assistance charge, and simultaneously reducing the price of toll calls, is that the telephone company enjoys no net gain or loss, while customers as a whole are better off by $8·35 M + 1·37 M = $9·72 M per annum. These calculations are set out in Table 9.1.

It should be noted that, although customers as a whole benefit by nearly $10 M per annum, those customers who make many local directory-assistance calls and few toll calls could be worse off, while those who make few directory assistance calls and many toll calls will almost certainly be better off. The economist cannot conclude that charges definitely ought to be charged for, without introducing some further value judgment. If, as appears to be the case, the largest users of directory-assistance services are firms wishing to make a quick credit check (Stannard, 1978, p. 98), most people, once they understand the issues and alternatives, will probably find the charge acceptable.

Table 9.1   *Analysis of DA charges and simultaneous reduction in toll price*

	Item	Company	Customers
1	Effect of DA charge on remaining DA calls	+ $10M	- $10M
2	Valuation of abandoned DA calls	+ $12M	- $3·65M
3	Effect of reduced toll price on original toll calls	- $22·5M	- $22·5M
4	Valuation of additional toll calls	- $ 0·5M	- $ 0·87M
	Totals	$0	$ 9·72M

Chapter 10

Peak-load pricing of telephone calls

10.1 Introduction

All telephone systems experience a distinct peak of traffic, usually during daytime hours. As a result, the marginal cost of providing calls in the daytime is far higher than in the night time, because in the peak period (and particularly in the busy-hour) extra capacity has to be provided while in the offpeak period capacity lies unutilised. These differences in cost are reflected in different prices of calls by time-of-day telephone systems. Usually only two different pricing periods are introduced, but in the more developed countries three or even four periods may be utilised (Roos *et al.*, 1976, B.M. Mitchell 1978*a*). Exactly how should these prices be determined, how should capacity be determined, and how do present prices and capacities in major telephone networks correspond to these optimal ones? This chapter outlines the theory of peak-load pricing, as it has been developed within the theory of welfare economics, and then describes an application designed to evaluate the pricing policy of the Illinois Bell Telephone Co.

10.2 Theory of peak-load pricing

For ease of exposition, we shall make a number of simplifying assumptions, most of which will be relaxed in subsequent Sections. Consider the demand for telephone calls over a single route. Divide the 24-hour day into two equal periods, called Day and Night, and assume demand is homogeneous (i.e. at a constant level) within each period and unaffected by price in the other period. Suppose that a given amount of capacity is available in each period, which allows up to y^0 calls in each period at a specified grade of service. Assume the marginal traffic cost of each call is negligible. The number of customers is assumed to be given, independent of call prices, and consumption externalities and congestion costs are ignored. Fig. 10.1 illustrates this situation, where the Day demand curve D_d lies above and to the right of the Night curve D_n. Following the principles set out in the previous chapter, to maximise net value of output on this route, the price per call in the daytime should be set equal to the market-clearing price p_d. The price per call in the night time should be zero, for even at zero price night-time demand x_n does not exceed available capacity y^0.

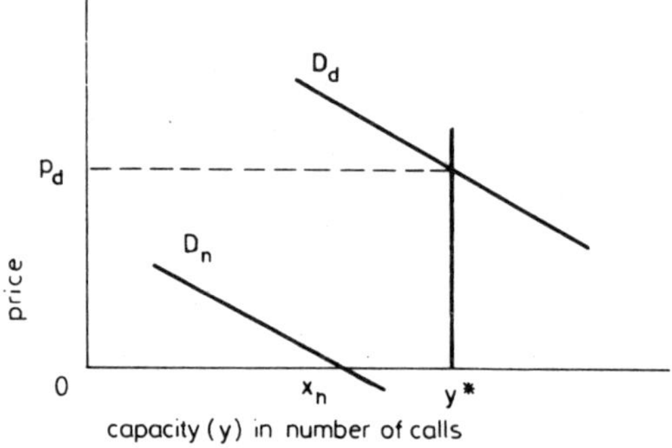

Fig. 10.1   *Peak-load pricing with fixed capacity*

Now suppose that capacity can be varied; should it be increased or decreased? At present, extra capacity at night has no value, and in the daytime has value p_d per call. Hence capacity should be increased or decreased according as p_d is greater or less than the marginal cost of capacity. If p_d is less than the marginal cost of capacity, capacity should be reduced and the daytime price increased so as to just clear the market. If capacity is reduced below the number of night-time calls x_n, it will be necessary to charge a positive price for each call in the night-time also. In this situation, prices are set so as precisely to flatten out the peak. Capacity has a value in the night as well as in the day. To obtain the total value of capacity, it will be necessary to add the marginal value in each period. In fact, a demand curve for capacity D may be obtained by *vertically* summing the Day and Night demand curves. Capacity should therefore be adjusted to the point y^* where the sum of the marginal values in the two periods (i.e. the sum of the market-clearing prices) is equal to the marginal cost of capacity. This situation is shown in Fig. 10.2, where marginal cost of capacity is assumed constant and equal to g.

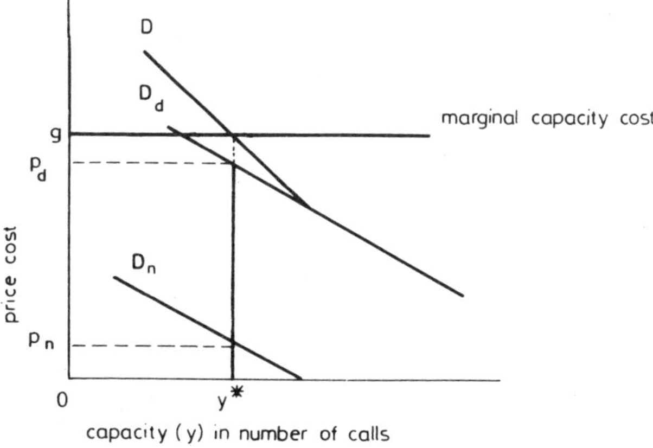

Fig. 10.2   *Peak-load pricing with variable capacity*

For those familiar with nonlinear programming models (see appendix to this chapter), the situation may be represented as follows. Let $p_1 (x_1)$ and $p_2 (x_2)$ be inverse demand curves in the Day and Night, respectively. The problem is to choose nonnegative outputs x_1, x_2 and capacity y to

maximise
$$\int_0^{x_1} p_1 (x'_1) \, dx'_1 + \int_0^{x_2} p_2 (x'_2) \, dx'_2 - gy \quad (10.1)$$

subject to
$$x_1 \leqslant y; x_2 \leqslant y$$

Let u_1, u_2 be shadow prices (dual variables) for the two constraints; u_i may be interpreted as the marginal value (or opportunity cost) of capacity in period i and is zero if there is spare capacity ($x_i < y$). The problem has the solution

$$p_1 (x_1) = u_1 , p_2 (x_2) = u_2 \quad\quad (10.2)$$

$$u_1 + u_2 = g \quad\quad (10.3)$$

The price in each period is set equal to the marginal opportunity cost of capacity in that period, which is zero if there is spare capacity, and the sum of prices equals the marginal cost of installing capacity.

This result may easily be generalised where there are many routes and many items of equipment, where demands during the day are interdependent, where there are many periods during the day, possibly of unequal length, and where marginal traffic costs are positive. At the optimum, the price for each route in each period equals the marginal traffic cost plus the marginal opportunity cost of capacity, and for each item of equipment the sum of the marginal opportunity costs of capacity equal the marginal cost of installing extra capacity of that item (Turvey, 1969, Littlchild, 1970c). The next section develops a model of such a network (nonmathematical readers may skip the following section).

10.3 Peak-load pricing in a network

Let there be calls over n different routes (subscripted j) during q different periods of the day (superscripted k) requiring capacity on m items of equipment (subscripted i). We shall assume a fixed routing pattern, so for each route j the set A_j of items of equipment used by every call on that route can be identified. Similarly, for each item i of equipment, the set B_i of routes using that equipment can also be identified.

There are two types of variables, both required to be nonnegative:

x_j^k = the mean volume of (3 min) calls per hour on route j during period k

y_i = effective capacity (in number of calls per hour at a specified grade of service) provided on equipment i

Since mean hourly demand varies within each period, we define the intraperiod peak factors

h_j^k = ratio of maximum demand (in any hour) for calls on route j in period k to mean demand (per hour) on that route in that period

The capacity constraints for each item of equipment in each period of the day are written

$$\sum_{j \in B_i} h_j^k x_j^k \leqslant y_i, \qquad i = 1, \ldots, m,\ k = 1, \ldots, q \qquad (10.4)$$

Let the hourly mean demand functions be denoted by

$$x_j^k(p^1, \ldots, p_j^q),$$

with the inverse functions

$$p_j^k(x_j^1, \ldots, x_j^q),$$

where the prices p_j^k are measured in cents per 3 min call. Let

t^k = duration (in hours) of period k, where $\sum_k t^k = 24$

c_j^k = constant marginal traffic cost (in cents per call) on route j in period k

g_i = constant marginal cost of equipment i (in cents per day per 3 min call capacity)

Note that we assume constant unit costs for each item of equipment. Economies of scale are reflected in the use of cheaper capital equipment on higher-volume routes. The capital costs include the allowed rate of return.

The maximand is the consumers' evaluation of calls made less traffic and capacity costs, or

$$\sum_j \left\{ \int \sum_k t^k p_j^k \, (\xi_j^1, \dots \xi_j^q) \, d\xi_j^k - \sum_k t^k c_j^k x_j^k \right\} - \sum_i g_i y_i \qquad (10.5)$$

where ξ_j^k is a variable of integration for x_j^k. The line integrals are taken from the origin to the points of optimal usage $(x_j^1, \dots, x_j^q)$ and are well-defined providing the integrability conditions

$$t^k \frac{\partial p_j^k}{\partial x_j^l} = t^l \frac{\partial p_j^l}{\partial x_j^k}, \qquad\qquad k \neq \ell, j = 1, \dots, n \qquad (10.6)$$

are satisfied (Pressman, 1970). It may be shown that these equalities constitute a reasonable approximation where only a small proportion of income is spent on the goods, as with telephones.

Let u_i^k be the dual variable corresponding to the typical constraint of expr. 10.4. It may be interpreted as the marginal opportunity cost (in cents per call) of using capacity i in period k. By complementary slackness, if an item of capacity is not fully utilised at any time, the marginal opportunity cost of using it is zero.

From the Kuhn – Tucker optimality conditions we derive the following characterisations of marginal-cost pricing and investment policy:

$$p_j^k = c_j^k + h_j^k \sum_{i \in A_j} u_i^k, \quad j = 1, \dots, n, k = 1, \dots, q \qquad (10.7)$$

$$\sum_k t^k u^k = g_i, \qquad i = 1, \dots, m \qquad (10.8)$$

According to eqn. 10.7, the price of a call in any period is set equal to its marginal traffic cost plus the sum of the marginal opportunity costs at that time of using the capacities it requires (allowing for the spare capacity necessitated by the intraperiod peak factor). According to eqn. 10.8 each item of equipment is purchased to the point where the marginal value of the capacity it provides, summed over all periods, equals its marginal cost.

10.4 An application of the model

This model has been applied by Littlechild (1970d) and Littlchild and Rousseau (1975) to a simple 3-route network in the Illinois Bell Telephone System, comprising a local call within Chicago, Ill. (route $j = 1$), an interstate toll call from Chicago to Peoria, Ill. ($j = 2$), and an interstate toll call from Chicago to New York, NY ($j = 3$). as illustrated in Fig. 10.3.

The day was divided into four periods, corresponding to those obtaining for pricing purposes in Illinois in 1967 (to which year all data refer), namely

Day	($k = 1$)	6 a.m. to 6 p.m.
Evening	($k = 2$)	6 p.m. to 8 p.m.
Night	($k = 3$)	8 p.m. to 12 p.m.
After midnight	($k = 4$)	12 p.m. to 6 a.m.

Six types of equipment were identified, then aggregated into five 'items'. The first three items contained transmission equipment particular to the three routes, the fourth item was toll-office equipment used by both toll routes and the fifth item contained outgoing exchange and local conduit equipment common to all routes. Current Bell System grades of service were adopted.

Traffic costs comprised wages of operators and relevant commercial and accounting staff, plus an allowance for uncollectibles (bad debts). Equipment costs consisted of capital or installation costs of the equipment, multiplied by an annualisation factor incorporating maintenance costs, property taxes, physical depreciation and obsolescence, and cost of capital (including interest, income tax, dividends and retained earnings).

Demand functions, assumed to be linear, were based on estimates obtained from Illinois Bell of traffic changes induced by Night-time rate reductions in 1962 and 1963. However, a number of additional assump-

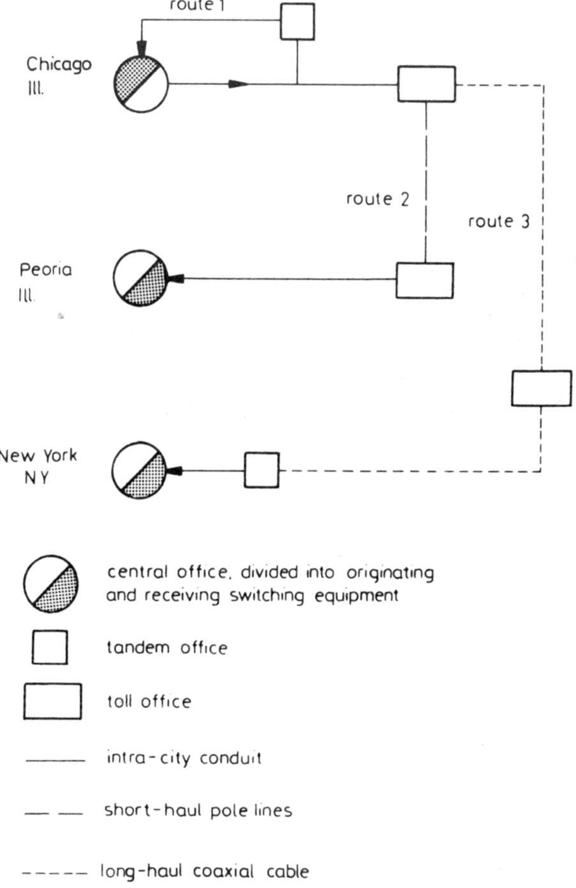

Fig. 10.3   *Example network*

tions had to be made to provide all the necessary crosselasticities of demand. For example, it was assumed that crosselasticity of demand between nonadjacent periods was zero. Extensive sensitivity analysis established that the results obtained were rather insensitive to the exact assumptions about elasticities.

Table 10.1 sets out the schedule of call prices obtaining in 1967, together with the schedules of prices generated by a variety of other pricing policies. Column 2 contains the schedule of prices equal to marginal cost, on the assumption that capacity is fixed at the level obtaining in 1967. It shows that, to utilise fully this existing capacity, the average level of prices would need to have been reduced by about

Table 10.1   Alternative pricing structures[a]

Route	Period	Actual (1967) situation	Marginal costs (fixed capacity)	Marginal costs	Profit maximisation	Ramsey prices	Ramsey prices (dual profit) constraint
Intra-Chicago (local call)	Day	5	5	3	10	6	5
	Evening	5	3	1	10	5	4
	Night	5	1	1	8	4	3
	After midnight	5	1	1	13	6	5
	Average[b]	5	3	2	10	5	4
	Daily volume of calls ('000)	7208	7195	8745	4372	6988	7301
Chicago–Peoria (Intrastate toll call)	Day	65	65	13	165	74	64
	Evening	50	34	5	194	81	67
	Night	40	5	5	141	60	50
	After midnight	40	5	5	112	48	40
	Average	54	38	9	168	73	62
	Daily volume of calls ('000)	99	101	120	60	96	100
Chicago–New York (interstate toll call)	Day	140	129	9	155	68	136
	Evening	100	95	16	113	55	100
	Night	70	60	29	73	47	68
	After midnight	70	5	5	74	33	65
	Average	105	93	17	116	57	103
	Daily volume of calls ('000)	107	118	194	97	155	109
Revenue[c]		549	466	235	671	543	544
Traffic cost		52	58	68	34	54	54
Operating profit		497	408	167	637	489	490
Capacity cost		141	141	167	83	133	135
Net profit		356	267	0	554	356	355
Consumer surplus		674	788	1107	277	707	684
Total surplus[d]		1030	1055	1107	831	1063	1039

a   All prices are in cents per 3 min call
b   Average price is calculated as the sum of prices in each period of the day weighted by usage in each period. Weighting by duration of period gives essentially the same result in most cases
c   These figures are in thousands of dollars per day. Subtractions may not exactly agree because of rounding error
d   Total surplus is the sum of consumer surplus and net profit (producer surplus)

38% for local calls, 30% for intrastate toll calls and 11% for interstate toll calls.

These price reductions would have generated traffic increases of about 10%, 2% and 10%, respectively, for the three types of route. With the capacity variable, as in column 3, marginal-cost prices would have been less than half 1967 price levels for local calls and about one-sixth 1967 levels for toll calls. Traffic increases of about 21%, 21% and 82% would have been generated.

The structure of prices exhibits, in most cases, a rather steeper decline from daytime to nighttime than at present obtains. Prices after midnight are reduced to the level of marginal traffic costs, because even at those levels demand is assumed to be less than capacity. When capacity is variable, it is optimal to expand it to such an extent, to meet daytime peak demand, that prices from 6 p.m. onwards fall to the level of marginal traffic costs for the local and intrastate routes.

On the long-distance interstate route a strange situation appears, for when capacity is variable, the marginal-cost price rises towards midnight. This is easily explained, but it highlights a difficulty with the model. At the 1967 level of capacity daytime callers have the highest marginal valuation of output, but their demand is least elastic because they are predominantly business users. As output increases the daytime demand curve falls fastest and in fact falls below the night demand curve at outputs in the vicinity of optimal capacity. At the same time, because of the linear approximation to the demand curves, elasticities fall to implausibly low levels. Linear approximations are probably inadequate when prices are cut by five-sixths and outputs are nearly doubled. If constant elasticity (log–linear) demand curves had been assumed, a more plausible price structure would probably have emerged, but because the demand curves would fall less slowly they would generate output increases perhaps up to tenfold. The basic difficulty is that US telephone systems have no experience of price cuts of the magnitude envisioned. On the interstate route, at least, the estimates of time-of-day price structure and resulting call volumes are not reliable (the *average* level of prices, however, depends on cost and is essentially independent of demand elasticities).

With fixed capacity, selling prices equal to marginal costs would have reduced the telephone company's profit by about one-quarter. However, consumer surplus would have increased by about one-sixth. The value of the telephone system as a whole (i.e. profit plus consumer surplus) would have increased by nearly 2·5%. With variable capacity, company profits would have been reduced to zero, but recall that these are 'excess' profits, over and above the allowed rate of return embodied in the capital costs. Consumer surplus would have been increased by two-thirds. The net effect would have been an increase in the value of the telephone system, to company and customers, of about 7·5%. If the model network were representative of the entire Illinois Bell network, this last figure would represent about \$60 M per year. Since Illinois Bell accounted for about 5% of all US telephone-company revenues in 1967, one might speculate that the adoption of marginal-cost pricing for telephone calls in the USA as a whole would have yielded net benefits of the order of one billion dollars per annum during the late 1960s.

10.5 Profit maximisation

Regulation obviously did not have the effect of equating the prices of telephone calls to marginal cost in the Illinois Bell network in 1967. What price structure would correspond to other pricing philosophies? Which pricing philosophy best explains the structure of prices actually obtaining on these routes in 1967? What light is shed on the effects of regulation? To answer these questions, the model may easily be modified to reflect alternative objective functions or additional constraints. For brevity, we report the results for variable capacity only.

Column 4 of Table 10.1 shows that, if the company had been maximising profit, prices of local calls would have been doubled and prices of intrastate toll calls would have been trebled. However, prices of interstate toll calls would hardly have been increased at all. In other words, the prices of telephone calls between Chicago and New York were effectively set at profit-maximising levels in 1967! These results suggest that the state regulatory commission was able to keep down the prices of intrastate calls, both local and toll, but that the FCC was quite unable to do the same for intrastate calls. Of course, prices of toll calls have been steadily reduced over the last decade, reflecting significant reductions in cost from new transmission technologies. However, it is well-known that several modifications in the 'separations procedures' used by the regulatory commissions to allocate common costs among the various services have had the effect of systematically transferring a larger proportion of costs from intrastate to interstate calls (Horn, 1977). We do not have comparable information for years other than 1967, but it seems conceivable that these modifications in separations procedures were designed precisely to shift as much of the burden of common costs onto the interstate calls as that traffic was capable of bearing, or at least, that the modifications had that effect.

Profit maximising prices would have led to an increase in company profit of some 60% over its actual level in 1967, but a decrease of the same percentage in consumer surplus. The net effect would have been a 20% decrease in the value of the system. For Illinois as a whole (if the model network is representative), these figures mean that unconstrained profit maximisation would have increased company profits by \$177 M at the expense of a decrease \$362 M in consumer surplus, a net loss of \$185 M per annum. It should be pointed out, of course, that in the absence of regulation it would be easier for new entrants to join the industry and the resulting competition in certain areas of business, notably heavily-trafficked long-distance routes, would presumably bring down prices below the monopoly levels assumed in the Table.

10.6 Ramsey prices

Either marginal-cost pricing or profit maximisation would have led to company profits substantially different from those actually achieved in 1967. By contrast 'Ramsey prices' (see chapter 9) are obtained by maximising net value of calls (or consumer plus producer surplus) subject to a minimum profit constraint, equal to profit actually obtained in 1967. They are shown in column 5 of Table 10.1. This philosophy would have required 1967 prices of intrastate toll calls to be increased by about one-third and prices of interstate toll calls to be cut by nearly half. Customers as a whole could have been made about 3% better off, without any reduction in telephone company profit.

It is difficult to compare these results with those of Willig and Bailey (1977) because the latter do not attach numerical figures to their distance axis. However, as far as one can judge, the results are not consistent. Littlechild and Rousseau calculated that Ramsey prices on the 1000 mile interstate route from Chicago to New York would differ from actual 1967 prices by the following percentages: Day, -51%, Evening, -45%, Night, -33%, After midnight, -53%. If we assume that this route lies about midway along the Willig and Bailey's short–long spectrum of milage bands, they calculated that the following percentage changes would be required to 1973 prices: Day, $+33\%$, Evening and Night, -20% (these figures are approximate, having been read off a graph). For log–linear rather than linear demand functions, their proposed changes are very different, namely: Day, -35%, Evening, -45%, Night, $+60\%$. In neither case, however, are they at all similar to those of Littlechild and Rousseau. Presumably, one explanation is that the set of routes included in the two models are very different.

Of course, not all customers would have been better off as a result of adopting Ramsey prices; those whose calls were predominantly intrastate toll calls would have been decidedly worse off. It is, therefore, of interest to examine the scope for making most customers better off by calculating Ramsey prices subject to a *pair* of profit constraints, one for interstate calls and the other for intrastate calls, again equal to the respective levels obtaining in 1967. This device ensures that not too much profit is recouped from intrastate calls. The resulting prices. shown in the last column of Table 10.1, are not at all dissimilar from the prices which actually obtained in 1967; the most noticeable change is, in fact, an average increase of under 15% in intrastate toll prices. Customers are about $1 \cdot 5\%$ better off overall; these savings would have been worth about \$8 M in Illinois as a whole in 1967. It is doubtful, however, whether in practice the savings could have been obtained; uncertainty about demand limits the practicable extent of 'fine-tuning' the tariff, and the cost of installing a metering system for local calls in Chicago alone was estimated at over \$15 M.

It is interesting to establish that 1967 pricing policy could be approximated by a philosophy of maximising net value of calls in the telephone system as a whole, subject to a pair of minimum profit constraints. It might be thought that this result is obvious, because the pricing policy is essentially determined by the pair of profit constraints.This is not so: maximising profit, gross revenue or total number of calls subject to the same two constraints leads to quite different pricing structures.

The result also highlights the roles of the state and federal regulatory commissions in determining the relative levels of interstate and intrastate profit. A comparison of the Ramsey prices with and without the second profit constraint (given the level of total profits to be earned by the company) suggests that the effect of state regulation is to reduce prices of intrastate calls, both local and toll, by about 15% but almost to double the price of interstate calls. The net cost of this state influence in terms of lost consumer surplus amounts to nearly 4·5% of 1967 revenue. For Illinois as a whole this would amount to about $21 M per year. The company's annual loss of $35 M profit on intrastate business is exactly compensated by increased profit on interstate business. However, allocating the equipment costs to routes on the basis of 1967 usage reveals that the net loss of $21 M on consumer surplus is actually a gain of $46 M to intrastate callers offset by a loss of $67 M to interstate callers. This redistribution of benefits to the state commission's 'constituents' presumably justifies its existence. If this calculation is correct, it represents counterevidence to the conjecture of Stigler and Friedland (1962) that regulation is ineffective, but is consistent with Stigler's (1971) later theory of economic regulation. We return to this matter in the final chapter.

Appendix   Some elementary results in nonlinear programming

Consider the problem of choosing the values of nonnegative variables $x_1, \ldots, x_n$ to maximise the value of a continuous and concave function $f(x_1, \ldots, x_n)$, called the *objective function*, subject to the chosen values of the variables satisfying a set of *constraints*

$$g_1(x_1, \ldots, x_n) \leqslant b_1$$
$$\cdots \cdots \cdots \cdots$$
$$g_m(x_1, \ldots x_n) \leqslant b_m$$

where the constraint functions $g_i(x_1, \ldots x_n)$ for $i = 1, \ldots m$, are continuous and convex, and the right-hand-side terms $b_1, \ldots, b_m$ are positive constants. Letting x denote the vector $(x_1, \ldots, x_m)$, the problem may be written

$$\max_x f(x)$$

subject to $g_i(x) \leq b_i, i = 1, \ldots, m$ (10.9)
$$x_j \geq 0, \; j = 1, \ldots, n$$

This is the typical form of a *nonlinear programming* problem.

It was shown by Kuhn and Tucker that a necessary and sufficient condition for $x_1 = (x_1^*, \ldots, x_n^*)$ to be an optimal solution to expr. 10.9 is that there exists a nonnegative vector $u = (u_1, \ldots, u_m)$ of so-called *dual variables* (or *shadow prices*) such that the following *dual constraints* or *duality conditions* hold:

$$\sum_{i=1}^{m} u_i \frac{\partial g_i}{\partial x_j}(x^*) \geq \frac{\partial f}{\partial x_j}(x^*) \, , j = 1, \ldots, n \qquad (10.10)$$

$$u_i \geq 0 \qquad\qquad , i = 1, \ldots, m$$

In a sense, these conditions are an extension of the 1st-order optimality conditions for an unconstrained maximisation, in which the first derivatives are set equal to zero.

The following complementary slackness conditions also hold for an optimum solution:

$$u_i \left[b_i - g_i(x^*) \right] = 0, \quad i = 1, \ldots, m \qquad (10.11)$$

$$x^* \left[\sum_{i=1}^{m} u_i \frac{\partial g_i}{\partial x_j}(x^*) - \frac{\partial f}{\partial x_j}(x^*) \right] = 0, \quad j = 1, \ldots, n$$

The first condition in eqn. 10.11 requires that either $u_i = 0$ or $[b_i - g_i(x^*)] = 0$ (or both). Equivalently, if the ith constraint is *slack*, that is, $g_i(x^*) < b_i$, then its corresponding shadow price u_i must be zero, whereas if the *ith* shadow price u_i is strictly positive then its corresponding constraint must be *tight*, that is, $g_i(x^*) = b_i$.

The second complementary slackness condition in eqn. 10.11 may be interpreted similarly. The conditions of exprs. 10.10 and 10.11 are known as the *Kuhn–Tucker conditions*. An excellent exposition of nonlinear programming from the point of view of the economist is given by Baumol (1972, chap. 7).

Chapter 11

Grade of service

11.1 Current practice

Morgan (1976, pp. 116-117), introduces the concept of *grade of service* as follows:

> ... it is not economical to provide switching equipment to carry the peaks of traffic orginated in the busy hour because the quantity required to cater for these fluctuations of traffic would be very large and the over-all time for which all would be occupied would be small. To avoid the uneconomical provision of equipment and to ensure that the average occupancy time of the equipment is high it is arranged that some calls are allowed to fail through insufficiency of switches. This permissible loss is known as the *grade of service* and may be expressed as the percentage of calls lost or as the proportion of time during which all the selecting devices at a particular stage in the connection are engaged.

This definition is the standard one. It has the unfortunate and confusing (to the layman) consequence that a low grade of service is a good one (for the user) whereas as high grade of service is a poor one.

Morgan does not explain further how the term 'economical' is to be interpreted, nor how to distinguish in practice between an economical and an uneconomical provision of equipment.

In practice, according to Syski (1960, p. 18), 'calculations are usually made on the so-called "fixed grade of service principle" '. That is, a (busy-hour) grade of service is fixed for the network as a whole, or for its constituent parts, and plant is engineered (or dimensioned) accordingly.

This approach has two major defects. The first is that it takes no account of the fact that large groups of circuits (or switches) are not only more efficient than small groups, but are also more sensitive to congestion (i.e.. their grade of service deteriorates by a higher proportion for a given percentage increase in traffic). In general, therefore, it will be more efficient to engineer different stages of the network for different grades of service, rather than impose a single uniform grade of service throughout. In practice, some account is taken of sensitivity to congestion by specifying two grades of service: that for the forecast (or measured) traffic, and a higher figure for a given overload (e.g. 10%).

The second defect is that the fixed grade of service principle does not, in itself, specify what that grade of service should be. Short (1976, p. 37) attributes the following view to Mina (1974):

> There is no rationale for the absolute level of existing grade of service standards beyond the fact that they have been found in practice to contribute to a workable quality of service within a reasonable price range.

Once again, there is no explanation of what constitutes a 'workable quality of service' or a 'reasonable price range'. Short develops his argument further:

> The present [grade of service] standards are widely assumed to have originated in the early days of automatic telephony . . . However, the standards have changed very little despite changes in the economics of the system. Although the standards have been apparently justified in the sense that they have proved workable over many years of operating experience, they cannot now be said to be rational in the sense of being based on any theoretical formulae or any known form of optimisation between the subscribers' satisfaction and the economics of the system (even if this were ever the case).

There have in fact been several economic (or 'rational') approaches put forward. The one best-known to telephone engineers is 'Moe's principle', dating back to 1923, which attempts to maximise the value of the system to the telephone administration. Welfare economists have developed analogous principles which attempt to maximise the value of the system to both administration and subscribers. These ideas actually predate Moe's principles, but it is only in the last decade that they have been applied to telephone systems. This chapter will describe and appraise the nature and implications of these various approaches.

11.2 Moe's principle

A rational economic approach to determining the optimal grade of service was put forward by K.O. Moe in 1923. It has been expounded in English by Brockmeyer and Jensen (1950), and a brief account is given by Syski (1960, pp. 648-54). Syski notes that in 1924 Erlang considered the rational determination of switch quantities in compliance with Moe's principle. He also mentions papers by Rodenburg (1949), de Kroes (1952, 1954) and Capello and Sanneris (1956) as important applications of Moe's principle.

Consider the problem of determining the optimal number of trunks to install on a single route with given traffic offered to it. Assume for the present that calls meeting fully engaged capacity are lost forever rather than held in a queue or repeated later. Suppose it is desired to maximise net revenue (profit). *Moe's principle* states that the optimal number of trunks is such that the revenue from an additional trunk is equal to the cost of providing it. Evidently it is a specific instance of the economist's principle of choosing output or capacity to equate marginal revenue and marginal cost (see chapter 8).

This result is easy to prove. Let

A = volume of traffic offered to a single route (in erlangs)

x = number of trunks installed

$E_x(A)$ = average proportion of lost calls on a route with A erlangs traffic offered to x trunks

p = average revenue per erlang

c = marginal cost per trunk (per hour)

If the number of trunks is increased from x to $x + 1$, the additional traffic carried is

$$F_x = A[1 - E_{x+1}(A)] - A[1 - E_x(A)] = A[E_x(A) - E_{x+1}(A)] \quad (11.1)$$

It will be profitable to add a trunk if and only if $pF_x > c$; hence at the optimum

$$\frac{p}{c} F_{x-1} \geqslant 1 \geqslant \frac{p}{c} F_x \quad (11.2)$$

or, equivalently,

$$pA[E_{x-1} - E_x] \geq c \geq pA[E_x - E_{x+1}] \quad (11.3)$$

Alternatively, if we treat the number of trunks as a continuous variable and write $E(A,x)$ instead of $E_x(A)$ for the grade of service, the problem is to

$$\underset{x}{\text{maximise}} \; pA[1 - E(A,x)] - cx \quad (11.4)$$

The solution may be obtained by differentiating to obtain

$$- p A \frac{dE}{dx} = c \tag{11.5}$$

At the optimum, the value of the marginal reduction in lost traffic just equals the marginal cost of capacity. Optimal grade of service is thus determined implicitly, as a result of choosing the optimal amount of capacity.

Moe's principle is easily extended to a network. Suppose there are n exchanges, and let

> x_{ij} = number of trunks linking exchanges i and j, $i, j = 1, \ldots, n$
> y_i = number of outgoing switches at exchange i
> z_j = number of incoming switches at exchange j.

Given the volumes of traffic offered between each exchange, we may write the amounts of traffic carried between each exchange as functions $\phi(x_{ij}, y_i, z_j)$ of the numbers of trunks and switches installed. Now let

> p_{ij} = average revenue per erlang carried between exchanges i and j
> c_{ij} = marginal cost per trunk between i and j
> g_i = marginal cost per outgoing switch at i
> h_j = marginal cost per incoming switch at j

The problem is to

$$\underset{x, y, z}{\text{maximise}} \quad \sum_{i=1}^{n} \sum_{j=1}^{n} p_{ij} \, \phi_{ij} \, (x_{ij}, y_i, z_j) \tag{11.6}$$

$$- \sum_{i=1}^{n} \sum_{j=1}^{n} c_{ij} x_{ij} - \sum_{i=1}^{n} g_i y_i - \sum_{j=1}^{n} h_j z_j$$

The solution is given by the following equations:

$$p_{ij} \frac{\partial \phi_{ij}}{\partial x_{ij}} = c_{ij}, \quad i, j = 1, \ldots, n \tag{11.7}$$

$$\sum_{j=1}^{n} p_{ij} \frac{\partial \phi_{ij}}{\partial y_i} = g_i, \quad i = 1, \ldots, n \tag{11.8}$$

$$\sum_{i=1}^{n} p_{ij} \frac{\partial \phi_{ij}}{\partial z_j} = h_j, \quad j = 1, \ldots, n \tag{11.9}$$

Each item of equipment is installed at the level at which the value of a marginal increase in traffic equals its marginal cost. Note that, whereas each additional trunk increases traffic between the two exchanges which it links, each additional switch at an exchange increases traffic to or from all other exchanges.

11.3 An appraisal of Moe's principle

Moe's principle is an application of economic principles to telephone engineering. Instead of adopting a fixed grade of service as a rule of thumb, the optimal grade of service is derived in relation to the objective of the business. This objective may be the maximisation of profit (as illustrated in the previous Section), but it could alternatively be some other objective more appropriate to a regulated or nationalised industry. Whatever the objective, a value is attached to lost or repeated calls, which is weighed against the cost of avoiding them. Moe's principle could also be applied if overall grade of service were specified for each route or type of call, and it were required to determine grades of service at each individual stage of the network so as to minimise total cost.

It is noteworthy that Moe's principle emphasises *improvements* in the grade of service, rather than the *absolute level* of grade of service. Moe defined the *improvement function* of each item of equipment as the rate at which its grade of service improves with respect to an increase in its capacity, multiplied by the volume of traffic offered to that item. This is the term $-A\ dE/dx$ in eqn. 11.4. Moe's principle thus states that the improvement function for each item of equipment is proportional to its marginal cost.

It follows from Moe's principle that the optimal grade of service will vary between different stages of the network, and between different networks, depending on volume of traffic offered, relationship between capacity and grade of service, average revenue of lost traffic, and marginal cost of increasing capacity. Specifically, it will be optimal to provide a better grade of service (i) the higher the volume of traffic offered, (ii) the greater the rate of improvement in grade of service as capacity is increased, (iii) the higher the average revenue on lost traffic and (iv) the lower the marginal cost of capacity.

One apparently serious drawback to Moe's principle is often mentioned (Newstead, 1961, Short, 1976, p. 31). Unless congestion is very severe, calls are rarely lost forever (i.e. abandoned), but are merely repeated until successful. If, in the limit, *all* failed calls were repeated until successful, there would be no loss of revenue from a poor grade of service. Moe's principle would provide no guide to provisioning, or, more precisely, it would seem to recommend installing just sufficient capacity to meet all traffic offered (eventually) at a 100% grade of service! More generally, it would seem that, insofar as calls are repeated, revenue alone provides insufficient incentive to provide an adequate grade of service.

Is this objection applicable to the British telephone system? Assuming that 50% of failed calls are repeated, Short (1976) estimates that congestion caused a maximum loss of telephone revenue of £6·2 M in 1974/75. This would seem to be an overestimate insofar as he cites evidence to suggest that 60–70% of failed call attempts resulted in a repeat attempt, and further that the liklihood of a subsequent repeat attempt actually increased after the failure of the first repeat attempt (see also Myskja and Walmann, 1971). Short remarks that this is 'not in itself sufficient to justify any improvement in quality of service'. Yet he has previously calculated that providing equipment merely to handle the existing congestion would cost about £3 M per annum. If this cost estimate were reliable, an improvement in grade of service would certainly seem to be indicated, even though the revenue estimate is likely to be an overestimate. In other words, contrary to the objection raised, Moe's principle based on net revenue would seem to indicate a better, rather than worse, grade of service in Britain.

It must be admitted, however, that Short's calculation of cost rests on some dubious assumptions, notably

(i)   that 'peak congestion *can be* twice the mean figure for the day as a whole' [italics added]
(ii)   that congestion will be eliminated by increasing capacity by a percentage equal to the peak grade of service thus calculated
(iii)   that equipment costs are a linear function of capacity
(iv)   that call revenues are an estimate of the cost of providing the equipment.

It would seem interesting and worth while to carry out a more thorough calculation based on estimates of costs and effects on congestion made by engineers in the administration itself (if, indeed, such a calculation has not already been carried out).

Another objection to Moe's principle from the opposing direction may also be noted. On many routes in many countries, prices are surely greatly in excess of marginal costs. Is it not possible that Moe's principle based on net revenue will lead to a grade of service which is too good? Some companies are alleged to provide their own internal telephone systems because they prefer a cheaper system, albeit at a worse grade of service, than the telephone administration offers. In the last two chapters of this book we shall discover a reason why a regulated telephone company might wish to offer a more expensive and better grade of service than customers would really prefer.

Common to these two opposing objections is the feeling that the telephone administration does, or should, look beyond its own interests (in revenue) to the interest of its customers and society as a whole. Specifically, it should take some account of the fact that a poor grade of service annoys subscribers and wastes their time.

An analogous criticism has, until recently, been applicable to queuing theory as a whole, not merely in telephone systems. One French critic begins his paper thus (freely translated):

> Queuing theory is one of the major branches of operational research, and its specialists are constantly advising on the practical operation of various diverse activities. Practically, all this advice is contrary to the social interest because it fails to take account of the social cost of waiting. [Kolm, 1970*a*]

A further objection to Moe's principle is that it takes no account of the feedback from grade of service (in the past) to volume of traffic offered (in the future). Perhaps this is a legitimate approximation for advanced telephone systems, where grade of service is extremely low, so that congestion is a negligible deterrent. But this is certainly not the case in the majority of developing countries, where congestion is often very severe.

A final objection to Moe's principle is that the volume of traffic offered is assumed to be determined exogenously. Yet there is increasing evidence (surveyed in chapter 3) of the extent to which traffic offered responds to price. In effect, a higher price is an alternative way of improving grade of service.

What seems to be required, then, is a more general principle than Moe's, which (i) takes into account costs and benefits to customers as well as to the telephone administration, (ii) chooses prices as well as capacities and (iii) takes into account the feedback of prices and grade of service on the volume of traffic offered.

Welfare economists have addressed themselves to precisely these kinds of issues for the past half century. To be sure, only recently have a few economists explored the implications for grade of service in telecommunications systems (Marchand, 1967, 1973, Kay, 1968, Pyatt, 1972). However, the problem of traffic congestion on roads has been much analysed (Pigou, 1920, Knight, 1924, Walters, 1961, Smeed 1964, Levy-Lambert, 1968) and recently the same ideas have been applied to canals (Lave and De Salvo, 1968) and airports (Carlin and Park, 1970). There is a related, but more general, literature on optimal pricing when demand is stochastic, i.e. uncertain, but following a regular and predictable pattern (Boiteux, 1951, Dreze, 1964). Finally, operational researchers have begun to examine the possibilities of 'social optimisation' in queuing systems (Naor, 1969, Kolm, 1970a,b, Yechiali, 1971, Littlechild, 1974).

In the remainder of this chapter we shall attempt to illustrate these ideas with a few simple models. It will be convenient to begin with a telephone link containing a single trunk in which there are facilities for holding calls at times of congestion until the line is free. Demand is assumed to be homogeneous (i.e. having the same probability distribution) throughout the day. The task is to choose the optimal grade of service, which is determined indirectly by choosing the optimal price per call and thereby influencing the volume of traffic offered to the single trunk. We then show how this model may be generalised to allow for multiple trunks, and how the optimal capacity (number of trunks) is determined. The model may be extended to incorporate varying demands at different times of day. An alternative model is indicated to handle automatic telephone systems in which calls are not held, so that failed calls are lost. Finally, there are some remarks on various other matters concerning grade of service which arise in practice.

Before discussing these analytical models, it will be useful to discuss in some detail the nature and extent of congestion costs.

11.4 Costs of congestion

The form and extent of traffic congestion, and its consequences for the subscriber, depends on the type of equipment in use and the policy adopted by the telephone administration:

(*a*)   In most automatic exchanges, shortage of capacity is reflected in the inability to obtain a dial tone, or in a signal that all circuits between certain exchanges along the desired route are busy. In such cases, it is necessary to redial until the call is successfully completed.

(*b*)   Some automatic exchanges have the facility to hold calls, perhaps up to a maximum number, and present the stored calls in order as and when facilities become available. Where an exchange is manually operated, a subscriber has to wait until the operator accepts his call. At times of congestion operators are usually instructed to accept calls in order of dialling. How far this is achieved depends upon the operators' memory, care and judgment, so that an element of chance inevitably enters. In these exchanges the subscriber has to wait while holding the handset.

(*c*)   International calls frequently need to be made through an operator. If all the international lines are busy, the operator may be willing to try the call at an approximate time in the future. Alternatively, it may be possible to book a call in advance for a specified day and time, providing an application is made early enough. Here, the subscriber does not need to hold the handset while waiting, but must be at the 'phone near the appointed time.

Congestion thus imposes various costs on subscribers. One component is the cost of time spent redialling. Pyatt (1972) estimated that, if it took an average of 30 s to establish a call, using labour valued at an average £0·60 per hour, and if 10 000 M calls were made annually in Britain, total costs of establishing calls would be of the order of £ 50 M per annum. We may calculate that if, say, 1% of these calls needed to be redialled because of insufficient equipment, the cost of redialling caused by congestion would be of the order of £0·5 M.

Short (1976) provides rather more detailed calculations along these lines, also for Britain:

	Local calls	Trunk calls
Total successful calls (1974/5), M	13523	2313
Success rate, %	63·1	61·9
Total attempts, M	21450	3730
Congestion, %	0·4	1·5
Number of call attempts encountering congestion, M	86	56

Average holding time for calls encountering congestion was 27·1 seconds, hence total lost time due to congestion was

$$(86 + 56) \times 10^6 \times \frac{27\cdot1}{60 \times 60} \text{ hours}$$

which gives a total of approximately 1 million hours annually. Taking the hourly wage of the typical customer encountering congestion as about £3, Short calculates that the total value to customers of the time wasted in the face of congestion is about £3 M per annum. In the light of inflation and the growth of traffic since 1974, this figure could perhaps be doubled. Whether it is correct to value savings of a few seconds or minutes at the corresponding fraction of the hourly or daily wage has received some attention (Tyler, 1973), but the question is still controversial and empirical work on the subject is sorely needed.

The time spent redialling is, however, only one component of congestion cost. A second component is the time wasted *between* attempts to redial. If a customer decides to wait 1 min before redialling, and if he cannot do anything productive in that time, then the total time spent redialling is approximately 1·5 min per attempt, or three times the cost just calculated (see Clos and Wilkinson, 1952, on the dialling habits of telephone customers).

Delays in trying to reach the operator, while applicable to a minority of calls in a mainly automatic network, can nevertheless be substantial. As this chapter is being written, there are complaints in *The Times* that it is not uncommon to spend 20 min trying to reach the international operator from Oxford (where direct international dialling is not yet possible). If time is valued at £3 per hour, as in the paragraph above, the congestion cost is £1 per international call.

Regardless of dialling time, delay in completing a call also imposes costs. A subscriber may be willing to pay very highly indeed to avoid delay in receiving or transmitting news of an accident, a rebellion, a company's financial results, a poor harvest, a failure to catch a train etc. Alternatively, it may be expensive or inconvenient to begin making a call sufficiently early to counteract the likely congestion. It may even be necessary to call more than once. One particular device to avoid congestion, often alleged to be used in Britain, is to send a telex to an overseas contact requesting him to telephone back; apparently the cost of this arrangement is less than the cost of waiting to telephone.

Finally, more important than the loss of time, may be the frustration and annoyance generated by congestion (Palm, 1953). Even if dialling time is only 0·5 min, subscribers may be willing to pay handsomely to reduce this chore to a few seconds. The popularity of push-button (touch-tone) and automatic (punched-card) dialling facilities surely reflects this view. There have been attempts to model the annoyance associated with congestion (Palm, 1953), but as yet there are no quantitative studies.

These considerations suggest that the total costs of congestion in Britain may be much higher than the estimate of £3 M per annum presented by Short, perhaps an order of magnitude higher. Even allowing for the lower value of time, congestion costs must be extremely high indeed in developing countries, where equipment often operates near full capacity throughout the day. In Costa Rica in 1973, for example, the load curve for certain exchanges was practically constant, equal to maximum capacity, from 8.00 a.m. to 10.30 a.m. and again from 1.00 p.m. to 4.00 p.m. In India even worse examples have been reported with delays of hours or even days to complete calls over some routes.

Although most telephone administrations seem to have a fairly good idea of the extent of congestion (as reflected in grade of service) at various parts of their network, there seem to have been very few attempts to measure the costs which this congestion imposes on their subscribers. None of the papers referenced by Short (1976) seems to go beyond the illustrative calculations presented above (work on this topic is understood to be in process in the British Post Office, and perhaps elsewhere). In view of the likely costs of congestion, the absence of empirical work is greatly to be regretted.

Admittedly, empirical work would not be easy. Subscribers very seldom have the opportunity to choose the grade of service they require, in the light of different prices, and thereby implicitly to express the value which they attach to (avoiding) congestion. However, one of the attractions of a private telephone system within a country, or of a tied line, is that it can provide a grade of service significantly different from that offered in the public network, at a correspondingly different price. It may be possible to learn something about the value of avoiding congestion by studying, say, the conditions under which a dedicated line is purchased, the number of such lines purchased, the level of traffic put on the lines and the resulting grade of service chosen etc. The growing literature on measuring shortage costs in electricity supply systems may provide further ideas (see Webb, 1977, Munasinghe, 1977, and the references therein).

11.5 A single trunk with calls held during congestion

Consider a single trunk line with facilities for holding in a queue however many calls arrive during times of congestion. It is assumed that no calls are abandoned before completion, and that calls held in the queue are taken in order of arrival as soon as the line is free. Assume further that the mean rate of arrival of calls is λ calls per hour (it will be convenient to use standard notation from queuing theory). Holding time (i.e. duration of call once connected) is assumed to be exponentially distributed with mean duration $1/\mu$ hours per call. For the system to attain equillbrium, it is necessary that the hourly arrival rate λ be less than the maximum hourly 'capacity' μ, otherwise the queue of calls held would gradually build up without limit.

This is the simplest model in queuing theory, and its properties are well-known (Syski 1960). The grade of service, denoted a, is the probability of finding the line busy, which is

$$a = \lambda/\mu \tag{11.10}$$

Evidently, grade of service varies directly (linearly) with mean arrival rate calls and with mean holding time.

The average number of calls in the system, including those being held in the queue and that being made, denoted $N(\lambda)$, is

$$N(\lambda) = \frac{\lambda}{\mu-\lambda} \tag{11.11}$$

As mean arrival rate λ increases to maximum capacity μ, the expected number of calls in the system increases to infinity.

The average time required to complete a call, including waiting time and holding time, denoted $T(\lambda)$, is

$$T(\lambda) = \frac{1}{\mu-\lambda} \text{ hours} \tag{11.12}$$

This, too, increases to infinity as mean arrival rate tends to capacity.

Suppose that subscribers value time at an average of £w per hour; this includes both time cost and the nuisance value of waiting. The average cost per call to a subscriber, excluding the price of the call, is then

$$£\frac{w}{\mu-\lambda}$$

We shall call this the average congestion cost of a call. When congestion is negligible, it is merely the value of the holding time w/μ. Evidently, it tends to infinity as the mean arrival rate of calls increases to the capacity of the single circuit. Total congestion costs incurred by all subscribers, denoted $G(\lambda)$, are

$$G(\lambda) = £\frac{w\lambda}{\mu-\lambda} \text{ per hour} \tag{11.13}$$

Assume that the mean hourly arrival rate of calls varies inversely with the average cost per call, where the latter equals price plus expected value of time spent. Write the inverse demand function $f(\lambda)$. It is assumed to be downward sloping, and measures the marginal value of a call as a (declining) function of the number of calls already offered. The gross value of all calls offered, denoted $V(\lambda)$, is measured by the integral of the inverse demand function

$$V(\lambda) = \int_0^\lambda f(\lambda')\,d\lambda' \qquad (11.14)$$

Finally, assume that the cost to the telephone administration of operating this system is an increasing function $C(\lambda)$ of the mean arrival rate of traffic.

We now explore the implications of choosing the mean arrival rate of calls so as to maximise the net value of the system to 'society' as a whole, i.e. the gross value to subscribers of calls made less the operating costs of the administration less the congestion costs to subscribers. Revenues from calls are merely transfer payments from subscribers to the administration, and may for the moment be omitted. Formally, the problem is to choose λ to

$$\text{maximise} \quad \left\{ V(\lambda) - C(\lambda) - G(\lambda) \right\} \qquad (11.15)$$

$$\text{i.e.} \quad \text{maximise} \quad \int_0^\lambda f(\lambda')\,d\lambda' - C(\lambda) - \frac{w\lambda}{\mu-\lambda}$$

The optimal solution is obtained by differentiating (11.15) to obtain

$$f(\lambda^*) = \frac{dC}{d\lambda} + \frac{w}{\mu-\lambda^*} + \frac{w\lambda^*}{(\mu-\lambda^*)^2} \qquad (11.16)$$

At the optimum, the marginal value of a call should just equal the marginal operating cost of a call, plus the average congestion cost of that call, plus a third component to be explained.

The optimal price £p^* per call is that which generates the socially optimal calling rate. Subscribers will make calls to the point where their marginal value equals their marginal cost, i.e. price plus congestion cost, or

$$f(\lambda) = p + \frac{w}{\mu-\lambda} \qquad (11.17)$$

Equating eqns. 11.16 and 11.17, we deduce that the optimal price should be set equal to marginal operating cost plus the third component of eqn. 11.16:

$$p^* = \frac{dC}{d\lambda} + \frac{w}{\mu-\lambda^*} \; \frac{\lambda^*}{\mu-\lambda^*} \tag{11.18}$$

This optimal price may be given a simple explanation in terms of 'externalities'. When a subscriber decides to make a call, he takes into account its cost to him alone. He ignores the fact that his call probably delays the completion times of several calls following his, thereby imposing congestion costs on other subscribers. In other words, the marginal cost of a call to him is less than the expected marginal cost of his call to all subscribers taken as a whole. Marginal *private* cost is said to be less than marginal *social* cost. However, to achieve a welfare optimum, it is necessary to face each subscriber with marginal social cost. It is, therefore, necessary to charge a price per call equal to the divergence between marginal private and marginal social cost. In this case, the required amount is the expected cost of the delay per call (equal to the expected time to complete the call $T(\lambda)$ valued at w) multiplied by the expected number of calls delayed $N(\lambda)$ plus the marginal operating cost incurred by the telephone administration. Thus eqn. 11.18 may be written

$$p^* = \frac{dC}{d\lambda} + w \, T(\lambda) \, N(\lambda) \tag{11.18a}$$

The grade of service may seem to have disappeared from the problem; in fact, given that there is one single channel, the optimal grade of service a^* is determined implicitly by optimal arrival rate of calls

$$a^* = \lambda^*/\mu \tag{11.19}$$

where this optimal arrival rate is, in turn, determined by optimal price per call.

To sum up, the prescription of welfare economics is as follows. Suppose it is desired to maximise the net social value of the existing telephone system (consisting of a single trunk in the present analysis). Net social value is defined as the aggregate money value of the system to all subscribers, less the cost of operating the system. In this case, the price of a telephone call should be set above the marginal operating cost of that call by an amount equal to the cost of the *extra* congestion which that call imposes on other subscribers. Grade of service is thereby determined at a level which is lower (i.e. better) than it would be if subscribers' congestion costs were ignored.

11.6 A numerical illustration

It will be helpful to illustrate these ideas numerically and graphically.
Suppose that average duration of a call is 3 min or 0·05 h, so that
maximum capacity is $\mu = 20$ calls per hour. Suppose that the value of
time is $w = £5$ per hour. Then the average congestion cost is $£5/(20-\lambda)$
per call. This cost function is shown in Fig. 11.1. Now suppose that the
inverse demand function is $f(\lambda) = 1 - 0·05\lambda$, which is also shown in the
same Figure. If a zero price is charged per call, it may be calculated that
mean calling rate will be 10 calls per hour. The grade of service is there
$10/20 = 0·5$, i.e. 50% of calls have to wait. The average number of calls
in the system is $10/(20 - 10) = 1$. The average time required to complete
a call is $1/(20 - 10) = 1/10$ hours or 6 min; of this, 3 min is spent waiting in the queue. Thus the average congestion cost is $£5/10 = £0·50$
per call. This, of course, coincides with the marginal value per call
$f(10) = 1 - 0·05\,(10) = £0·50$.

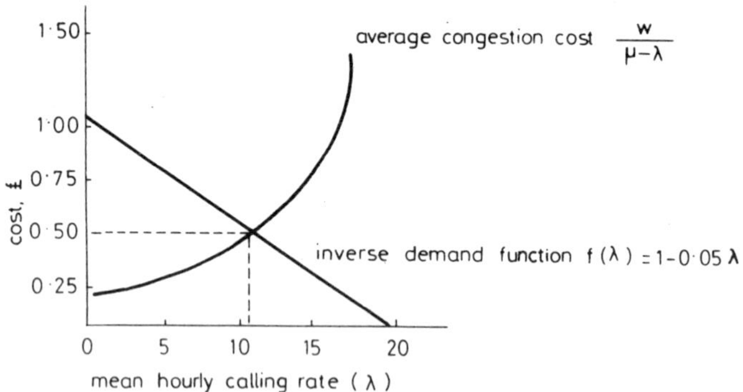

Fig. 11.1   *Congestion cost of calls*
$\mu = 20,\ w = £5$

A zero price and the resulting level of traffic and grade of service are
by no means optimal. The marginal cost to *all* subscribers of an additional call is given by the function

$$\frac{5}{20-\lambda} \cdot \frac{1+\lambda}{20-\lambda} = \frac{100}{(20-\lambda)^2}$$

This marginal social-cost curve is shown in Fig. 11.2. It rises more
steeply than the marginal private cost curve, which equals average
congestion cost. Assume for simplicity that operating costs are zero. At
a mean arrival rate of 10 calls per hour the marginal social cost of a call

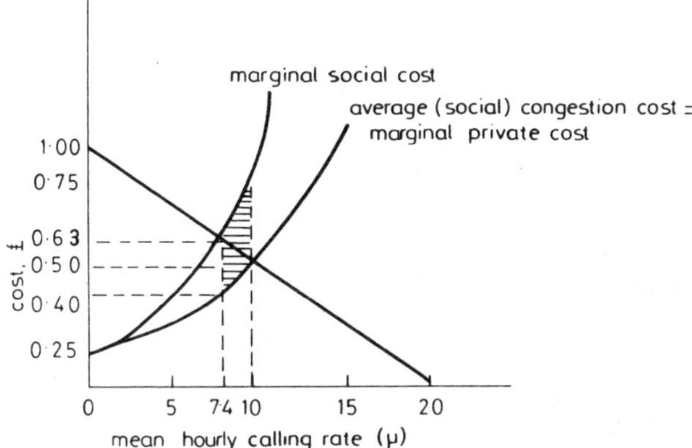

Fig. 11.2   *Determination of optimal calling rate (and implicitly grade of service)*

is £1, or twice the marginal private cost, and hence twice the marginal value of that call. The mean traffic volume should, therefore, be cut back to 7·4 calls per hour, at which rate marginal value equals marginal social cost equals £0·63 per call. Since at this traffic rate average congestion cost is £0·40, a price of £0·23 per call is required. The optimal grade of service is 7·4/20 = 0·37.

The value of the reduction in congestion is given graphically by the shaded area in Fig. 11.2. The loss in value of calls is

$$V(10) - V(7{\cdot}4) = \int_{7{\cdot}4}^{10} (1-0{\cdot}05\lambda)d\lambda = 2{\cdot}6 - \left.\frac{0{\cdot}05\,\lambda^2}{2}\right|_{7{\cdot}4}^{10}$$

$$= £1{\cdot}47 \text{ per hour}$$

whereas reduction in congestion costs is

$$G(10) - G(7{\cdot}4) = 10 \times £0{\cdot}5 - 7{\cdot}4 \times £0{\cdot}40 = £5 - £2{\cdot}95$$

$$= £2{\cdot}05 \text{ per hour}$$

The net saving is £2·05 − £1·47 = £0·58 per hour. Since the net value of the system with zero price was $V(10) - G(10) = £7{\cdot}50 - £5 = £2{\cdot}50$, this represents an increase in net value of nearly one-quarter. There is, however, a transfer of revenue from subscribers to the telephone administration, in the form of increased charges, in the amount of 7·4 × £0·23 = £1·70 per hour. If this revenue can in some way be returned to subscribers—for example, by reducing the monthly rental charges—then the average subscriber will be approximately 25% better off. This is a simple and clear illustration of how pricing according to the principles of welfare economics can improve consumers' welfare.

This philosophy has been explicitly adopted to solve the problem of traffic congestion in Singapore, apparently with great succes (World Bank, 1976). Since 1975 a tax has been imposed on vehicles entering the city during the morning rush hour. The resulting reduction in traffic has reduced peak-hour journey times by about one-third and has generated substantial revenues for the government. All concerned are reported to be satisfied with the outcome. It would seem likely that increasing prices of peak-period telephone calls would have a similar happy outcome in most developing countries where busy-hour grade of service is far too high.

11.7 Multiple trunks and variable capacity

The model of Sections 11.6 and 11.7 is equally applicable to a single route with $k \geq 1$ trunks available. Analytic expressions for grade of service, average waiting time etc. may still be derived (Syski, 1960), but they are no longer so tractable. It will be more convenient to work with general functional forms and diagrammatic illustrations.

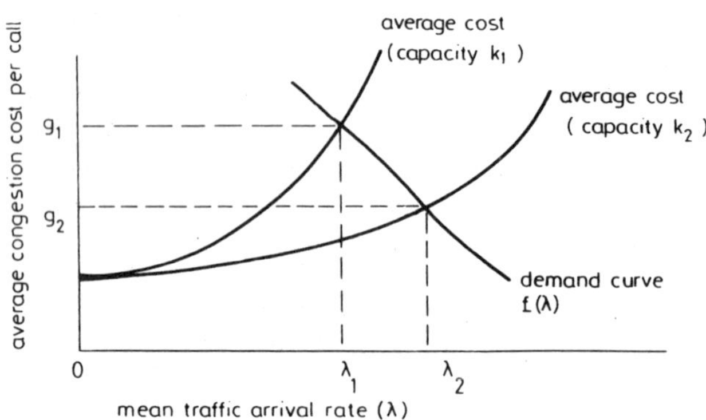

Fig. 11.3 *Effect of increasing capacity*

The effect of varying capacity may be represented graphically as in Fig. 11.3, which shows the average congestion cost per call as a function of traffic for each of two levels of capacity. Suppose price per call is zero. If capacity is increased from k_1 to k_2, the average congestion cost falls from g_1 to g_2 and the mean arrival rate of calls increases from

λ_1 to λ_2, even though price and the demand *curve* itself remain unchanged.

This simple diagram explains a phenomenon which appears to be common in many developing countries experiencing severe congestion. An increase in capacity is planned to accommodate existing traffic and eliminate congestion. Then the new capacity is installed, traffic increases to fill the capacity available, and the system is still congested! The reason is quite simple. A reduction in congestion is a reduction in cost and, effectively, a reduction in price from the subscriber's point of view. An expansion in traffic is only to be expected. To illustrate, it was mentioned earlier that subscribers in Oxford are currently experiencing delays of 20 min on international calls. Suppose the average price of such a call is £1·50 and the average value of time spent waiting is £4·50 per hour. In current congested conditions, the effective total price of the call is

$$1 \cdot 50 + \frac{4 \cdot 50}{3} = £3$$

Eliminating congestion would amount to a 50% cut in price (equivalently, the introduction of congestion effectively doubled the previous price).

Let us proceed to characterise optimal policy in the general case. As before, let $1/\mu$ denote average holding time, λ average arrival rate of calls, $f(\lambda)$ the inverse demand function, w the subscriber's average value of time and p the price of a call. Let $C(\lambda,k)$, $G(\lambda,k)$ and $T(\lambda,k)$ denote, respectively, total operating costs, total congestion costs and average time a call spends in the system, as functions of mean arrival rate and number of trunks (the latter assumed a continuous variable for analytic convenience). Let $H(k)$ denote the cost (per unit time) of providing k trunks. Continue to assume that all calls are held until the system can accept them.

The problem is to choose mean arrival rate and number of trunks so as to maximise net value of the system:

$$\text{maximise } V(\lambda) - C(\lambda,k) - G(\lambda,k) - H(k)$$

Writing $G(\lambda,k) = \lambda T(\lambda,k)$, we have

$$\text{maximise } \int_0^\lambda f(\lambda') \, d\lambda' - C(\lambda,k) - w \lambda T(\lambda,k) - H(k) \qquad (11.20)$$

By differentiation, necessary conditions for an optimal solution are

$$f(\lambda) = \frac{\partial C}{\partial \lambda} + wT + w\lambda \frac{\partial T}{\partial \lambda} \qquad (11.21)$$

and

$$\frac{\partial C}{\partial k} + w\lambda \frac{\partial T}{\partial k} + \frac{\partial H}{\partial k} = 0 \qquad (11.22)$$

Since the subscriber himself incurs the average congestion cost wT per call, eqn. 11.21 implies that price should be set equal to marginal operating cost plus marginal congestion cost imposed upon all other subscribers:

$$p = \frac{\partial C}{\partial \lambda} + w\lambda \frac{\partial T}{\partial \lambda} \qquad (11.23)$$

The new eqn. 11.22 says that capacity should be increased to the point where its marginal cost equals the value of the reductions in operating cost and congestion cost which it makes possible (increasing capacity is assumed to lower operating and congestion costs, so that the first two derivatives on the right-hand side of eqn. 11.23 are negative).

11.8 Peak-load pricing

In the peak-load pricing model discussed in the previous chapter, it was assumed that the 'fixed-grade-of-service principle' operated. For a given route, the price per call in each period of the day is set equal to the maximum of the marginal operating cost and the capacity clearing level, where capacity is defined in terms of this fixed grade of service. The resulting price structure leads to a complete flattening of the load curve over those (peak) periods where capacity clearing prices obtain, and over these (peak) periods grade of service is constant. In off peak periods grade of service is necessarily lower.

It should be recognised that, in practice, a completely flat load curve would be unattainable, because demand is not as elastic or predictable as the model assumes. The role of the theory is to provide a useful characterisation of an 'ideal' target. However, once congestion costs are introduced, the uniform grade of service and the flat load curve are no longer appropriate targets, even if demand could be manipulated with ease.

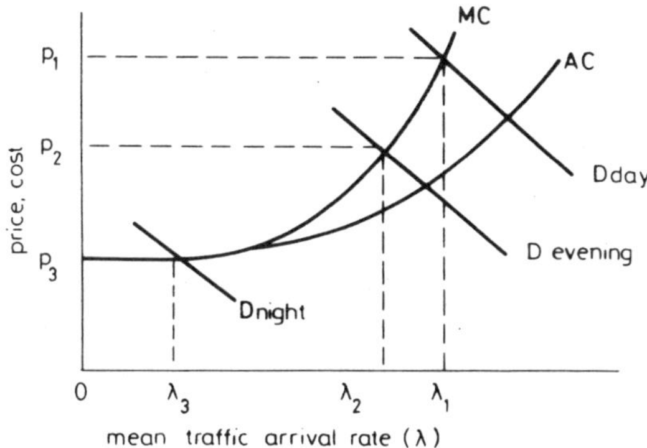

Fig. 11.4   *Peak-load pricing with congestion*

This may easily be demonstrated in Fig. 11.4, which shows the average (AC) and marginal (MC) congestion cost curves corresponding to an item of equipment with fixed capacity. Assume zero operating costs. Three different demand curves are shown, corresponding to Day (1), Evening (2) and Night (3) periods. Optimal prices and levels of traffic are also indicated, following the principles of Sections 11.6 and 11.7. It is apparent that mean traffic arrival rates are not equated by the optimal call prices, with the result that grades of service are not equated either. In the absence of congestion costs, peak demands are met solely by raising prices; when congestion does exist, peak demand is optimally met, in part, by allowing grade of service to deteriorate. Whereas the optimal load curve is *less* flat than with uniform grade of service, the optimal time-of-day structure of prices is *more* flat than with uniform grade of service.

Where capacity is taken to be variable, the optimal policy is comparable to that for peak-load pricing. Capacity should be extended to the point where its marginal cost equals its marginal value in reducing congestion and operating costs, where the latter savings are obtained by summing over all the periods of the day for which the equipment provides capacity. The extension to whole networks of equipment is straightforward, and need not be set out in detail here.

Peak-load pricing models with endogenous grade of service have been analysed by Kay (1968), Pyatt (1972), Marchand (1973), Deschamps

(1977). The latter has made some calculations for the prototype Illinois Bell network described in the last chapter. Grades of service are allowed to be variable, and determined within the model, rather than specified in advance. For the (evening) peak periods, prices of intrastate calls are approximately half what they would have been with uniform grade of service, and peak grades of service are correspondingly higher.

11.9   Congestion costs in networks without queuing facilities

We have hitherto assumed that calls arriving at times when the system is congested are stored in order of arrival until the system is free; in this case congestion costs are assumed to be related to average waiting time. In most telephone networks, such queuing facilities do not exist (with a few exceptions, as discussed earlier). Congestion is reflected in a busy signal. Congestion costs are related to the time lost in redialling and waiting between dialling attempts, and the corresponding annoyance. We may, therefore, assume that congestion costs are proportional to the grade of service. In effect, the queuing system has changed from 'first-come, first served' to random service.

In principle, two analytic approaches are available. The first, taken by the authors just referenced, is to specify demand curves for telephone calls as a function of price and grade of service in each period. These models explicitly determine optimal grades of service, and are often analytically convenient. However, they generally require the use of some form of 'social welfare function' which is not a natural approach for telephone engineers,and possibly less easy to apply. On the other hand, the alternative approach of maximising net value of output in the telephone system, which has been adopted earlier in this chapter, appears to be more complicated when queuing is not possible. To our knowledge, it has not been developed in any detail. We may indicate the source of the difficulty and the nature of the approach; it will be seen that the qualitative results are quite comparable to those of systems allowing queuing.

The difficulty which arises is that traffic *offered* is not the same as traffic *carried*. It is natural and usual to define grade of service, congestion and costs in terms of traffic offered, but it is more natural to define demand for calls, and value thereof, in terms of traffic carried. Some device must be used to relate the two. It will be convenient here to relate all magnitudes to traffic carried.

For a price of equipment of given capacity, plotting grade of service g against traffic offered x yields the S-shaped curve shown in Fig. 11.5.

In uncongested systems, only the first part of this curve is relevant, and traffic carried is approximately equal to traffic offered. But as congestion gets worse, the discrepancy increases. For example, if 10 erlangs of traffic is offered to 10 trunks, the grade of service is 0·215, i.e. over 20% of calls are lost, so that only 8 erlangs of traffic is actually carried. If traffic carried is plotted against traffic offered, a relationship of the form shown in Fig. 11.6 is obtained. The traffic carried gradually approaches, but cannot exceed, a theoretical 'full capacity'. If we now plot grade of service against traffic carried, the resulting curve is increasing throughout, as shown in Fig. 11.7. Taking congestion cost per call as proportional to grade of service on calls actually made thus yields average and marginal costs of congestion as shown in Fig. 11.8. These cost curves may then be used in exactly the same way as described in earlier sections. Optimal price per call is now defined in terms of the adverse effect on grade of service rather than on average waiting time.

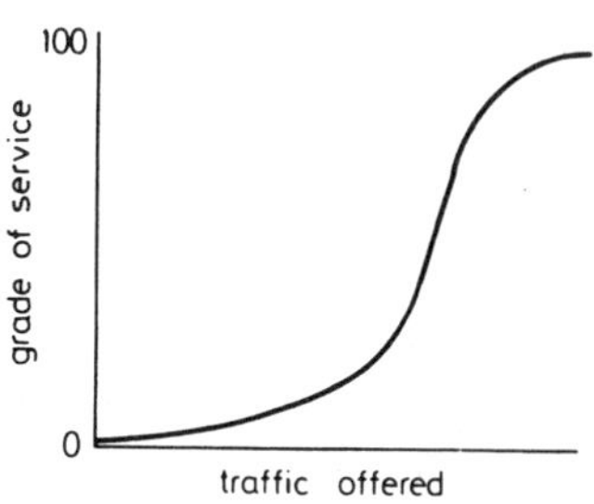

Fig. 11.5 *Grade of service as a function of traffic offered to equipment of given capacity*

Fig. 11.6 *Traffice carried as a function of traffic offered*

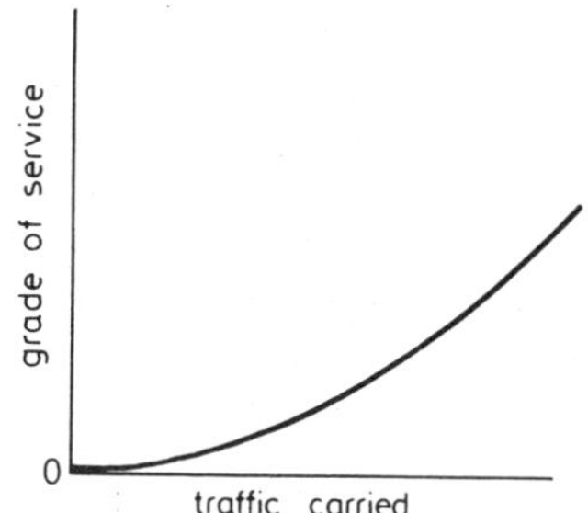

Fig. 11.7 *Grade of service as a function of traffic carried*

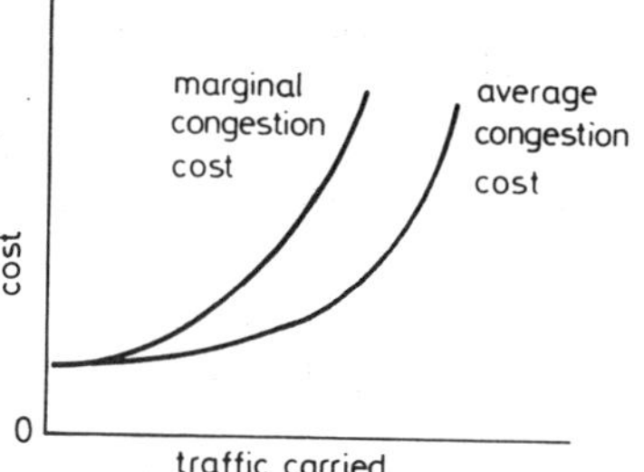

Fig. 11.8 *Congestion cost as a function of traffic carried*

Three final remarks are in order. We have hitherto assumed that duration of calls is independent of congestion and price. In fact, increased congestion is likely to increase the duration of calls. The reason is quite simple: waiting or attempting to make a 6- min call takes no longer than waiting to make a 3-min call. A 6-min call is effectively cheaper than two 3-min calls. Once the connection has been established, the cost of prolonging the conversation for the second 3 min is merely the price of a 3-min call: no further waiting or redialling is involved. The effect on duration of calls is likely to be greater, the greater is the extent of congestion. In some countries, indeed, it is found that customers tie up a line all day. This is certainly not socially optimal for the system as a whole, and is an additional argument for congestion pricing, which would remove the artificial incentive to prolong calls.

How far the methods of congestion pricing described in this chapter, are practical remains to be seen. In chapter 14 we discuss some of the limitations of welfare economics, which includes congestion pricing as a special case. For the present, we note briefly that it requires the telephone administration to look beyond its own financial books to the costs (often psychological) of its customers. This is a difficult and somewhat hypothetical exercise. Moreover, since it is difficult to calculate the socially optimal price, it is correspondingly difficult to enforce the policy on a recalcitrant administration. This leads us to examine what incentives the administration has to worry about customer costs. One such inventive is that subscribers will not be willing to pay high rental charges for a telephone system which is highly congested. Rental revenues thus provide some feedback (albeit indirect and imperfect) concerning subscribers' views about grade of service.

Finally, we have concentrated in this chapter on congestion and grade of service. In most telephone systems in practice, the number of calls lost due to the called subscribers' line being busy exceeds by at least an order of magnitude the number of lost by congestion in the public network. Thus, subscribers who rent too few lines or overload their PBX operators may be said to impose costs on people who call them! The logic of congestion pricing would seem to dictate that the charge for (additional) lines be set below the cost of supplying them so as to reduce the number of calls lost in this way to the socially optimal level. This problem does not seem to have been analysed by economists.

Chapter 12

Externalities

12.1 Introduction

The rules for optimal investment and marginal-cost pricing derived in the last few chapters are based on the assumption that there are no externalities in the telephone system. The costs and benefits incurred by managers and customers are assumed to constitute the costs and benefits incurred by society as a whole, i.e. private costs are assumed to equal social costs. In other words, the actions of any person (acting on behalf of himself or his organisation) either affect only himself (or his organisation), or else if they affect other people it is in ways which are known to the actor and taken into account by him.

In practice, these conditions are seldom met, and certainly not in telephone systems. It is, therefore, necessary to modify the rules to take externalities into account.

The previous chapter has already given one example. Each customer making a telephone call presumably considers the cost to himself, but ignores the additional congestion costs imposed on other callers. It was suggested that the telephone administration should adjust its prices of telephone calls so as to bring to bear on each customer the costs he imposes on others. His private cost will then equal social cost; his resulting decision whether to make the call will then be socially optimal.

In this chapter we explore various other aspects of externalities. The first springs from the widely held notion that each person joining the telephone network thereby confers benefits on other subscribers. Indeed, it has been suggested that each person originating a telephone call thereby confers on the person receiving the call. Second, market prices do not always correctly reflect the costs to society of supplying capital, labour and foreign exchange. In both of these cases, the principles of welfare economics require the telephone administration to modify its behaviour to attain a socially efficient outcome. Third, there may be irremovable constraints, beyond the jurisdiction of the telephone administration, which require what would otherwise be the best policy to be abandoned in favour of a 'second-best' policy. Finally, there may be equity considerations which require yet further modifications in behaviour to attain a 'socially optimal' outcome. To calculate such optimal behaviour in the face of so many complexities, it is often suggested that the techniques of cost-benefit analysis be used. We shall briefly survey the literature on these various topics.

12.2 Consumption externalities: views of US telephone administrators

The notion that access to a larger number of subscribers increases the value of the telephone system is a fairly obvious one. It probably dates from the beginning of exchange service. Much was made of this point during the debate, which took place in the USA during the 1890s, as to whether the hitherto prevalent practice of flat-rate charges should be replaced by measured rates. The absence of any charge on calls had so encouraged the 'telephone habit' by existing subscribers that, as Kingsbury (1915, p. 474) remarked,

> ... the effect of dividing the total outlay for the service by the total number of subscribers resulted in the establishment of a rate which was necessarily out of proportion to the expected use of a new subscriber.

This was especially felt in the large city. In its report of 1892 (p. 9), the American Bell Telephone Company observed, in defence of the flat-rate system, that

> ... in the large city containing some thousands of exchange stations, the use made by each subscriber and the value to him of the telephone facilities must inevitably be much greater than they would be in an exchange of less importance ...

Kingsbury commented on this passage as follows:

> that the larger the number of subscribers who were connected to the service, the greater was the advantage to all, needed no demonstration. Some system was clearly necessary which should be fair to existing users and not act as a bar to the introduction of new users.

By 1893 it was generally recognised that the flat-rate system did constitute such a bar to new subscribers, who were relatively small users. The subsequent message-rates were introduced from 1894 onwards, in order 'to bring the service within the reach of the numerous class of those who have need for a limited use of the system' (American Bell Telephone Company's Report, 1897, p. 8). An extremely rapid increase in number of subscribers resulted (Kingsbury, 1915, Fig. 170, p. 477).

Although reference was made to the value of the system depending on the number of subscribers, this concept of externalities does not appear to have been the real explanation for the change in the method of charging. A method was required which would allow the charge to a new subscriber more closely to reflect the marginal cost of serving him, regardless of any contribution he made to the other subscribers. No proposal was made to set charges to any subscribers below cost.

The concept of externalities (though not so-called) was much more central to the philosophy of 'universal service' subsequently propounded by Theodore Vail, president of the Bell System.

> The policy of the Bell system is that the value of a telephone service is in direct proportion to its 'unversality' and dependability'; that is, to the certainty of reaching promptly by telephone the greatest number of people . . .

> The Bell System makes rates for such kinds or classes of service as may be desired by, and will be acceptable to, each and every possible user. In this way it has made it possible for, and to the advantage of, every person to be connected with the exchange system who would add to the value of the service to others.
> [Vail, 1917, quoted in Horn, 1977]

Vail's concept of designing rate structures to achieve universal service was affirmed and ratified repeatedly in the decades that followed; by 1957, three quarters of the households in the USA had telephone service. Once again, however, it would seem possible to explain a part of these rate structures as devices to increase profit (e.g. by discrimination), which would hold even in the absence of externalities. In certain circumstances, it may even be profitable to 'subsidise' the entry of subscribers, by charging a subscription rate less than marginal cost, because the profit on the calls which these subscribers make outweighs the 'loss' on the rental. (Littlechild, 1975*b*). The real significance of externalities is thus difficult to evaluate.

12.3 Consumption externalities: analyses by economists

The first explicit, albeit brief, discussion of externalities in telephone systems is by Hazlewood (1950). He recognises the significance of the policies just discussed in the light of the existence of externalities:

> Every new subscriber increases the number of possible connexions, and so enhances the value of the service . . . . That is why particular importance must be placed on the establishment of 'promotional' tariffs, and why a tariff which, by charging more than cost, unnecessarily restricts the number of subscribers is so much to be deplored.

Hazlewood then goes further, though somewhat cautiously:

> The consideration of external benefits does suggest, however, that a strictly cost basis might with advantage be departed from in some circumstances, less than the full customer cost being charged to some subscribers whose connexion to the system is thought to yield a particular external benefit.

To clarify his proposal that charges to some subscribers be set below cost, Hazlewood appends a footnote:

> Such a policy is only reasonable if it is assumed, as is fairly clearly justified, that some subscribers enhance the general utility of the service more than others. If all subscribers were of equal value from this point of view there would be no grounds for departure from a cost basis.

This claim is not entirely correct. Even if all potential subscribers would enhance the general utility of the system by the same extent—for example, if its utility depended solely on the *number* of subscribers, and not their *composition*—there would still be a case for setting the subscription rate below the marginal cost. Of course, the financing of this policy then becomes a problem. It would only be possible to charge *all* subscribers a rate below marginal cost if a government subsidy were available, or if surplus revenues were available from profits on calls or subscriber equipment.

Analytic models of consumption externalities in telephone systems were developed by Littlechild (1970*a*, 1975*b*,), Gravelle (1972), Pyatt (1972), Squire (1973) and Turvey (1974). Somewhat different approaches, in the spirit of game theory, were provided by Littlechild (1970*b*), Artle and Arverous (1973) and Rohlfs (1974).

Without going into the details of these models, the main assumptions and results can easily be stated. Suppose that a subscriber's demand for calls depends on his income, the price of a call and the total number of subscribers. Suppose also that the telephone administration wishes to maximise the total net value of the system (i.e. consumers' plus producers' surplus). In the absence of any financial constraint and assuming that the price of each call is set equal to its marginal cost, the optimal policy requires that the number of subscribers be increased to the point where the marginal cost of accommodating a subscriber equals the marginal value which he receives *plus* the increase in value (i.e. consumer surplus) which he generates for all other subscribers. To attain this optimal number of subscribers, the subscription rate should be set *below* marginal customer cost by an amount equal to the external benefit received by other subscribers. This external benefit essentially comprises the consumer surplus enjoyed by existing subscribers on calls which can now be made to the new subscriber. It may be shown (Littlechild 1975*b*) that this is equal to the average consumer surplus per subscriber multiplied by the elasticity of demand for calls with respect to number of subscribers in the system.

An additional dimension to the externalities question has been provided by Squire (1973). He remarks that, of all public utilities, telecommunications 'affords one of the best instances of external economies. In fact, it would appear that an externality is created every time a phone is used because the recipient of the call obtains a benefit for which he does not pay'. The efficient allocation of resources thus requires that a telephone call be made if and only if the value of that call to the sender *plus* the value to the receiver exceeds its marginal cost. If one assumes, as a first approximation, that the expected benefit from an incoming call is the same for all subscribers, optimal policy is to set the price of a call below its marginal cost by the (constant) amount of this benefit.

12.4 Evaluation of externalities associated with calls

If the proposal to subsidise telephone calls to correct for external benefits is to be taken seriously, several conditions must be satisfied. First, the external benefits obtained by the recipients of calls must be substantial. Second, there must be a substantial number of calls which are *not* presently made but which 'ought' to be made if external benefits are taken into account. The fact that recipients of calls *presently made* derive great benefits from them is irrelevant. Third, there must be no alternative method available and worth using by which the potential recipient can ensure that the calls in question are made. Finally, subsidising telephone calls should not cause other distortions which more than offset the benefits of the subsidy. We shall briefly expand on these points.

Many incoming telephone calls undoubtedly bring pleasure to the recipient, notably those from family and friends. Not all do. Nuisance calls, by definition, are quite unwanted; the recipient would be willing to pay to *avoid* them. Since people typically have no direct way of purchasing incoming telephone calls, in the way that they can purchase a loaf of bread or a suit, it is difficult to infer their valuations from their behaviour.

There are, however, several ways in which the potential recipient of a call can express his preferences. If he suspects that the benefits from a particular call would exceed its price, he can originate the call himself or offer to pay for it. If he suspects that the combined value to originator and sender will exceed its price, though the value to either one does not, he can offer to share the expense. If a series of such calls is envisaged, the parties concerned can agree to take turns in calling. Even without making any financial payment, the potential recipient can make known his views to the potential caller, so that the latter is induced to take the former's preferences into account. Thus, it is common to telephone one's parents or children because one knows they would appreciate a call. Telephone (and postal) administrations nowadays emphasise such considerations in their advertising slogans — for example, 'Make someone happy with a telephone call' or 'Someone, somewhere, would like a letter from you'.

Where the potential caller is not known to the recipient, as with businesses seeking clients, it is possible to accept collect (reversed-charge) calls, or to use and advertise a 'toll-free' number; such devices enable the recipient to judge, at relatively low cost, whether further conversation is worth while.

All these devices are used regularly. It seems unlikely that there are many calls of any significance whose combined value to the two parties exceeds the price of the call, but which nevertheless are not made. In this case, an indiscriminate subsidy on *all* calls—and it is difficult to see how the telephone administration could identify separately those calls which *need* a subsidy—is likely to encourage many 'undeserving' calls without necessarily ensuring that all the 'deserving' calls are made. The end result is likely to be a much *less* efficient allocation of resources.

It should be made clear that the argument of this Section is entirely concerned with pricing to achieve efficiency, in the sense of welfare economics. It does not preclude the subsidising of calls by particular groups of subscribers (e.g. aged or sick people) who are thought to be particularly deserving from a social point of view, or whose use of the telephone it is thought desirable to encourage. This point is developed at greater length in a later Section of this chapter.

12.5 Evaluation of externalities associated with subscribers

The same kind of comments and qualifications can be made with respect to externalities generated by new subscribers. Certainly a telephone system consisting of a single subscriber is worthless, and certainly any addition to the network, by increasing the set of people it is possible to call or receive calls from, makes the network more valuable. But, once more, the question to ask is whether there are many people who do not have telephones, but who nevertheless ought to be in the network because enough existing subscribers would together receive benefits in excess of the marginal cost of adding them. Here, too, there are possibilities of internalising the externality. Subscribers themselves can install phones in their children's or their parents' apartments, or they can let other potential subscribers know that they would appreciate being able to call them. Doctors and bookmakers, who might not make any calls themselves, find that they have a direct financial interest in acquiring telephones so that clients can call them. If there *are* people who ought to be in the network but are not, the further questions arise—who are these potential subscribers, and how can they be brought into the network without also bringing in the 'wrong' people (i.e. those whose value to others is less than marginal cost)?

In the case of telephone calls, it seems likely that externalities could easily be internalised because there are only two parties to the typical telephone call. Negotiation is comparatively simple. By contrast, there are likely to be many potential beneficiaries of a potential subscriber, who may not all be known to each other; even if they were, it would be difficult for them to reach an agreement about sharing the rental fee. Thus, 'transactions costs' may prevent subscriber externalities from being internalised.

Where externalities are significant, is there any way of reducing the rental to these potential subscribers without also reducing it to others? It would seem possible selectively to reduce rentals or installation charges in specified geographical areas (where this is permitted by law), but there is no strong reason to believe that this would isolate the desired subscribers. The alternative to selective reductions is an overall reduction in rental, or at least a reduction for certain broad classes of customers (e.g. residential). Unless a government subsidy is to be provided, this will need to be financed either by increasing the rental to other groups of subscribers, or by increasing call charges. Both of these devices will cause their own misallocation of resources. These disadvantages must be balanced against the advantages of extending the network. Perhaps the advantages of reducing rentals are greater in networks in developing countries, where telephone penetration is presently low. In higher-income countries, where telephone penetration is already high, the case seems somewhat weaker.

12.6 Empirical evidence on externalities

Despite the popularity of the consumption externalities hypothesis there seems to be virtually no empirical evidence to support it. We shall briefly examine three predictions.

In an unpublished paper, Borts (1964) argued that

> the larger the community the more potential contact and messages among individuals, the larger the market, and consequently, the greater the choice of individuals to call. Boil it down and my hypothesis is that the demand for telephone service is a function of community size. The larger the community, the larger the expected number of telephones per household.

From examination of various New England data, Borts found that there were large differences in telephone development among metropolitan areas which were not explained by differences in income. He concluded that 'some part of these differences appears to be explainable by differences in the size of the metropolitan areas'. It is possible, of course, that the higher demand in larger communities could be explained by other factors, such as higher costs of communication by alternative methods. Moreover, it is not clear that a subscriber's circle of personal acquaintances will be significantly larger in a larger town, so that once his acquaintances are in the system, further subscribers represent little additional value to him. Indeed, in the study of telecommunications in Costa Rica, the present author found that the elasticity of connections with respect to size of population in the exchange was 0·8, hence number of connections per head *decreased* with population (see chapter 6). This study allowed for distance of the exchange from provincial and national capitals, but other variables such as income, which may have been relevant, were not included.

Externalities constitute the basis for one explanation of the S-shaped growth curve. Morgan (1976, p. 184) describes the so-called law of 'natural growth' as follows:

> In a community with only a few telephones the inducement to take up the service is much smaller than when there are many and it is argued that the number of existing subscribers will therefore exert an influence on the remaining non-subscribers to join the system. Suppose there are P potential subscribers at a given time 't' years when S subscribers and $(P - S)$ non-subscribers exist. The influence kS will be exerted on $(P - S)$ non-subscribers so that annual rate of conversion dS/dt will be given by kS $(P - S)$. That is the equation to the law is $(dS/dt) = kS (P - S)$.

The solution to this differential equation has three forms: logarithmic, hyperbolic and exponential. The latter may be written

$$S = \frac{P}{1 + \exp\left[-kP\,(t+c)\right]} \tag{12.1}$$

which is evidently a *logistic curve* of the form of eqn. 3.3 discussed earlier. On reflection, this is not surprising. If growth is proportional to outstanding nonsubscribers $P - S$, for constant population P it is necessarily proportional to number of subscribers S. This does mean, however, that an empirical test will not immediately distinguish between the 'externalities' hypothesis and a 'saturation' hypothesis. As remarked earlier (Section 3.3), one test of the hypothesis found that a constant saturation level P was inadequate to model the growth observed in the USA. It is possible that the saturation level may reflect other factors which are changing over time.

Finally, it would seem that if a subscriber gained greater value from a larger exchange, this would be reflected in the larger number of calls he made. One would expect, therefore, that average number of calls per subscriber would be higher in a larger exchange. To test this properly, it would be preferable to use data from the days when local exchanges were not interconnected. Kingsbury (1915, p. 468) reproduces certain figures from the National Telephone Exchange Association Report of 1881. Exchanges were grouped into four categories by size of city. Relevant data are as follows:

Group	1st	2nd	3rd	4th
Average population	400 000	83 683	19 680	5778
" number of subscribers	1,080	478	130	39
" price per year	$80·00	$50·00	$41·16	$38·40
" number of calls per	165	105	137(?)	131(?)

These data do not suggest a significant positive relationship between exchange size and number of calls per head. However, the report emphasises that 'the record must be accepted with some allowance' because some entries were made up by guesswork, especially in the record of calls per month. Another significant reservation is that new subscribers are comparatively small users, and existing (older) subscribers are comparatively large users. To compare the average calling rate over *all* subscribers will, therefore, understate the effect on the calling rate of the existing subscribers, which in princple is what should be measured.

The casual evidence cited here is not sufficiently reliable to disprove the externalities hypothesis, but neither does it offer support for it. An explicit attempt to test the hypothesis would seem to be indicated.

There must be considerable recent evidence that larger exchanges have higher calling rates, but this does not test the hypothesis because of the large proportion of junction calls. Certainly orders placed by administrations or manufacturers specify higher calling rates for larger exchanges. However, this is because they are for town centres with a higher proportion of business subscribers.

12.7 Shadow prices in a developing country

In most developed economies, market prices give a reasonable indication of the relative valuations which people attach to different products. In many developing countries this is not so, because the market mechanism is not well-developed. The possible reasons for this are many: the economies may be based to a much larger extent on self subsistence, incomes may be relatively low so that market demand is limited, information may be difficult to disperse so that prices are unstable over time, the system of private property rights may not be well-defined and enforced, and governments may impose various constraints on the market to modify its effects.

If a telephone administration is to make its major investment decisions in a way which uses the country's resources efficiently, it must recognise that 'market' prices in a developing country may have these severe limitations, and may not be a reasonable guide to relative scarcities. Economists have developed the concept of *shadow prices*, meaning the prices which reflect relative scarcity values of goods and services. These are the prices which, it is suggested, *would* obtain if the market were free to operate efficiently. One task of the telecommunications economist is, therefore, to identify divergences between actual prices and shadow prices, to estimate the size of these divergences, and to trace through the implications for pricing and investment appraisal (Dasgupta and Pearce, 1972, Little and Mirrlees, 1974, Squire and van der Tak, 1975, Little and Scott, 1976, Irvin, 1978).

Three major sources of divergences may be illustrated from the case of Costa Rica, discussed in chapter 6.

(*a*) The official *exchange rate* specified by the government may over value the domestic currency. This makes foreign imports appear cheaper than they really are, but exports appear less attractive. At the official rate of exchange, there is a greater demand for foreign currency (to import goods) than there is supply of it (from exports). Foreign currency has to be rationed; it acquires a scarcity value which must be taken into account when comparing the cost of imported equipment with that of domestic equipment. In Costa Rica in 1973, the official rate of exchange was 6·65 colones to the US dollar, while the free market rate was 8·50 colones. For purposes of social calculation, the purchase of telecommunications equipment should be valued at the free-market rate, which, of course, makes it more expensive. If this is not done, an excessive amount of importing will take place.

(*b*)   The telephone organisation, usually being government owned, may be able to obtain *financial capital* at a preferred rate, i.e. at a rate lower that that paid by private borrowers. In Costa Rica, it was found that the telephone company was able to borrow at about 7%–9% interest rate, while savings and loans companies paid 12% for deposits and charged from 15%–40% for loans. This preferred rate of interest will divert investment away from private use into less profitable government use, and it will also cause capital-intensive methods of production to appear less expensive than labour-intensive methods.

(*c*)   There may be a *minimum wage* in effect, which must be paid to unskilled labour, even if the labourers would otherwise be unemployed. In such cases the financial cost overestimates the opportunity cost of labour to society and again makes labour-intensive methods less attractive than they would be in a free market. In Costa Rica, despite the relatively low level of unemployment at the time of the case study, some economists estimated that the shadow wage level would be about 8%–9% below the current minimum wage.

If these three divergences between private and social costs are at all common throughout the developing world, they have a consistent implication for telecommunications. Calculations based solely on financial factors will bias decisions in favour of modern imported transmission and exchange equipment such as microwave, coaxial cable, automatic crossbar and electronic exchanges, and against more elementary equipment such as pole wire and manually operated exchanges. These countries may be led to modernise their step-by-step exchanges too quickly, thereby depriving their other industries of scarce capital and foreign exchange.

It is worth noting here that similar considerations can apply in developed countries. For example, Great Britain for many years maintained an artificially high exchange rate, which was accompanied by official exhortations to export and save imports. The Post Office is currently under great pressure not to cut modernisation programmes since this will cause losses of jobs in the telecommunications supply industry. In effect, it is being argued that, although the supplying firms do not have much alternative business, the wages of the employees either *will not* fall to a point where they would be employed elsewhere (because of trade union pressure) or *should not* fall to such a level (because that would be socially undesirable). The Post Office should, therefore, maintain its investment programme *as if* it were based on a much lower shadow wage, but, nevertheless, the current high wage should still be paid.

On the other hand, it might be argued that shadow prices are necessarily conjectures, and often unreliable. Once a financial basis is abandoned, it is not difficult to calculate shadow prices (or, indeed, to bring forward social considerations) which support whatever case one wishes to make. Some would argue that attention might better be directed at improving the operation of the market, to obviate the need for the calculation of shadow prices.

12.8 Second-best considerations

The theory of welfare economics characterises the pricing and investment policies which should be implemented throughout the economy to attain socially efficient and optimal allocations of resources. However, it may be shown that, if not *all* the conditions for efficiency can be satisfied, it will not necessarily improve matters to achieve the efficiency conditions in one particular industry. Each industry needs to take into account divergences from optimality elsewhere in the economy, particularly where goods which are close substitutes or complements are involved. The analysis of such situations, and the derivation of appropriately modified pricing and investment rules, is known as the *theory of the second-best* (Lipsey and Lancaster, 1956, Millward 1971).

The relation between postal and telecommunications business provides a good illustration. In most countries, it appears to be the case that the price of a letter reflects the average cost of delivery rather than the marginal cost; the latter is likely to be much lower. Deliveries in and between major cities are also likely to cost less than deliveries in rural areas though both have the same price equal to overall average cost. If, in contrast, prices of telephone calls were set to reflect marginal costs, customers would be faced with relative prices for letters and telephone calls which did not correctly reflect relative marginal costs. Insofar as some choice of mode of communications was available to them, customers would be overly biased towards telecommunications between cities and towards posts in rural areas. Resources would thus be misallocated. To avoid this, the theory of the second-best would require, instead, that the prices of urban and interurban telephone calls be increased above marginal cost, and the prices of rural calls be decreased (on the assumption that current letter prices must be taken as given). The actual extent of these modifications would depend on elasticities of demand and several other parameters. Similarly, investment plans for telecommunications should take into account the effect on the postal business.

Not all economists feel that second-best considerations are important. (Turvey, 1968*c*, pp. 86-9, Farrell, 1958). Common views are (i) that it is virtually impossible to take all such second-best considerations into account and (ii) that, if each industry takes the behaviour of the government and every other industry as given, improvement will be painfully slow. Rather, each industry should follow a first-best policy and attempt to convince other industries to do the same.

12.9 Cost-benefit analysis

It is evidently no simple matter for the telecommunications administration to look beyond its own account books to those of society as a whole. One way of systematically evaluating major investments is by the use of a *social cost-benefit analysis* (CBA). This is an attempt to set out the alternative actions available to the administration, to specify all the consequences of each action, to evaluate these consequences for society as a whole, and to choose the action which, on balance, seems most in the public interest. The usual procedure is that benefits and costs are evaluated according to the prices which consumers *would be prepared to pay*, given their initial incomes, rather than at the prices they actually *do* pay. The resulting calculations yield an *economic rate of return* to the proposed project. In general, this will differ from the *financial rate of return* because the latter will not reflect those costs and benefits to society which are not embodied in monetary payments.

Notable examples of cost-benefit analyses are the studies of the M1 motorway, the Victoria underground line and the third London airport (see the readings in Layard, 1972, and Munby, 1968). In the case of the M1 motorway, no revenues at all were envisaged from motorists, since tolls were not levied. Hence, benefits needed to be calculated with reference to time savings for motorway traffic. There were additional benefits for nonmotorway traffic, since traffic remaining on roads near the motorway experienced time savings from lower congestion as other traffic transferred to the motorway. In the case of the Victoria Line, the saving in traffic congestion on the roads above the underground was a most significant component. Other benefits included increased comfort on the underground. In the case of the airport, major considerations were the noise imposed on nearby residents, and the damage caused to the scenery. Since commodities such as time, comfort, noise, scenery are not directly traded on the market, attempts have to be made to infer peoples' valuation from their conduct, e.g. by comparing the sale price of a house near an airport with the price of a similar house in a quiet neighbourhood, or by observing the choices commuters make between slow, but cheap, modes of travel and fast, but expensive, ones.

Two difficulties arise with cost benefit analysis. First, the economic rate of return which results will generally (though not always) exceed the financial rate of return, because benefits are more likely than costs to have escaped payment. Private firms base their own calculations on financial returns. If they, too, were to use social calculations they would probably earn higher economic rates of return. Does this not lead to a distortion of investment in favour of public as against private projects? Is it possible to correct this bias by choosing an appropriate discount rate for the cost-benefit analysis? Specifically, if firms could earn 5% financial rate of return, what economic return would justify diverting capital away from the private firm? There is by no means agreement on this question.

Second, simply adding up willingness to pay may not yield an outcome agreeable to everyone's sense of fairness or national interest. Should fine scenery or unique historical monuments be given a greater value than people are prepared to pay for? If so, why? Does the use of willingness to pay mean that the preferences of rich people are being given more weight than those of poor people? If so, what alternative is there? The site for the third London airport recommended by the cost-benefit commission was, in fact, turned down by the government, presumably because of political pressure of one kind or another. This indicates that the view of the government and its supporters as to what was socially optimal differed considerably from the view of the Commission, which was based on net value of output to society as a whole, calculated with reference to estimated willingness to pay. Fieldstein (1972) has proposed a method for calculating 2-part tariffs in which greater social utility is attached to the income of poorer people. There is no doubt that this procedure reflects a popular demand by economists, but the economist, *qua* economist, is not able to specify the utility weightings which should be used, nor is there likely to be unanimous agreement among members of the public.

Chapter 13

Issues in regulation in the USA

13.1 Introduction

In about a dozen countries, the telecommunications system is operated by one or more private companies, subject to government regulation. In another score or so, there is some element of private capital involved, perhaps in partnership with the government, or in one or more designated sections of the telecommunications system. Such private ownership raises new issues.

In the previous three chapters we have explored the principles of welfare economics for telecommunications pricing and investment, on the assumption that the administration accepts these principles. Under this assumption, the chief task of economists is to elucidate the meaning and practical implications of the principles, rather than to enforce their adoption. Where the telecommunications system is operated by a regulated private company, this assumption is no longer valid. We discuss in the final chapter whether the assumption is valid even in a nationalised system. In a privately owned system a set of controls or guidelines must be designed which will lead the industry, under regulation, to an efficient and optimal outcome. In other words, the principles of welfare economics need to be adapted to the institutional environment.

The present chapter will discuss some of the policy issues that have arisen in the USA, and the economic analyses which have been developed to handle these situations. The same kinds of issues are likely to arise in all countries; they are particularly important in the USA because of the size of the telecommunications industry there, the rate at which technological progress raises new possibilities and the vigour of the private sector in attempting to exploit these opportunities, thereby repeatedly challenging the existing framework of regulation.

13.2 The framework of regulation in the USA

A very brief account of the concept of regulation in the USA will be helpful. This section is based upon Phillips (1969).

For many year, the Courts attempted to establish a separate category of businesses that were 'affected with a public interest' and that required detailed government regulation. Businesses so designated were generally privately owned, but known as *public utilities*; typical examples are transportation, power, water and communications. However, industries could be regulated without being classified as public utilities (e.g. the milk industry). Regulation often exists both at the federal level and within each state.

Regulated industries have acquired certain obligations and rights. The first major obligation is to serve all who apply for service, even if, at times, this is unprofitable. The second obligation is to render safe and adequate service; for the telephone industry this usually means service must be provided 24 hours a day, presumably without undue congestion. Third, the industries have an obligation to serve all customers on equal terms; unjust or undue price discrimination is forbidden, although customers may be classified (e.g. into business and residential subscribers) for purposes of rate making. Finally, the industries are obliged to charge only a 'just and reasonable' price for services rendered. The interpretation of this last provision has resulted in considerable controversy.

The regulated industries have the right to a reasonable price for their services–they cannot be required to operate at a loss–and they have the right to withdraw service under prescribed conditions. They often have the right of eminent domain; telephone companies, for example, may take land required for pole lines providing just compensation is paid. Most important, however, is that regulated industries have the right to *protection from competition* from enterprise offering the same service in the same area. Frequently, a franchise or other form of permission is necessary to operate and restrictions on the granting of franchises usually provide a significant measure of protection. This, also, is a controversial subject, particularly in telecommunications in recent years, since many new entrants have been clamouring for the right to compete.

Interstate and foreign telephone and telegraph services have been regulated by the Federal Communications Commission (FCC) since 1934. Intrastate telecommunications are regulated by state public service commissions established at different dates. The commissions are in theory independent of other branches of government, since the commissioners are appointed for definite terms of office. However, the commissions must work within guidelines specified by government, and regulation is inherently a political activity.

Most of the commissions' time, particularly at State level, is taken up by rate regulation. This has two aspects: control of the rate *level* (earnings) and control of the rate *structure* (prices). As to the former, the regulated industries are permitted to earn 'a fair rate of return on the fair value of their property'. This return must be neither so high as to exploit customers nor so low as to discourage investors. As to the rate structure, the industries are permitted to practice 'just and reasonable' price discrimination to expand the use of their services.

The major task of the commissions is thus to establish questions of fact rather than principle. What are (or were, or will be) the actual operating expenses of the utility? What is the net depreciated value of the utility's property, known as the *rate base*? What is a *fair rate of return* in the light of current, past and prospective interest rates? These three taken together determine the firm's *revenue requirement*. For purposes of approving rate structure, the commission must determine the costs of individual services, distinguishing between separable and common costs. It must predict the effect of rate changes on demands, on costs, and hence on prices to users of other services.

The FCC has been more concerned in recent years with the appropriate structure of the telecommunications industry; the role of competition, the entry of new firms, vertical integration between Bell and its supplier of equipment (Western Electric), policy on satellites and international cables etc.

13.3 Rate-of-return constraints

To ensure that customers are not exploited by a monopolist supplier, regulatory commissions have generally specified a maximum rate of return which may be earned on capital invested in the company. To make it worth while for the company to invest at all, this allowed rate of return should be set above the cost of capital. A margin above cost of capital is generally thought to provide some added incentive to discover new business and introduce new technology, since if the resulting profits or cost savings were automatically taken away the company would not find it worth while to pursue such lines of activity and exploration. Various schemes have, therefore, been proposed for allowing, say, half the profits in excess of the stipulated rate of return to be retained by the company. Others have argued that the central need is to provide incentives for managers rather than shareholders (see the discussion by Kahn 1971, vol. II, pp. 59–63, and the references therein, notably Morgan, 1923).

In this connection, economists have examined the phenomenon of *regulatory lag* (Trebing, 1963, Klevorick, 1966, Baumol, 1967, Kahn, 1968, Bailey, 1973). Regulated companies are not, in practice, continuously under supervision (despite the FCC's proposal of 'continuous surveillance'). Prices and costs are examined at intervals of several years, and if necessary the company is ordered to readjust prices to yield a fair rate of return. In this case, it would appear that there is an opportunity to earn excess profits, or to benefit from cost reductions, in the periods between examination. The effect of regulatory lag is thus two-fold; it reduces the control of monopoly power, but *ipso facto* provides the company with temporary incentives comparable to those facing an unregulated company. In this respect, incidentally, the relationship between the FCC and the operating companies is identical to that which existed between the British Post Office and its suppliers under the former Bulk Contracts agreement; see Section 14.8.

The most important limitation of the rate-of-return constraint, in the eyes of many economists, is that it distorts the decisions made by the company in various important areas. The first explicit analysis of this possibility by Averch and Johnson (1962) resulted in the use of the term *Averch–Johnson effect* (or *A–J effect*) to describe the bias generated by rate-of-return regulation. Briefly, the argument is as follows (see Baumol and Klevorick, 1970, Bailey 1973 and Kahn, 1970/71):

Suppose a company produces a single output using two inputs, say labour and capital, each available in unlimited quantities at constant prices. The company attempts to maximise profit but is limited to a specified rate of return r on its rate base, a return which is greater than the cost of capital c, but less than the return which the company would earn if not regulated. For every unit of capital which the company raises, it pays the market cost c, but gains an increase in take-home profit $r-c$, so that the *net* cost of capital is effectively $c - (r-c)$. For example, if the cost of capital is 8% and the allowed rate of return is 10%, the net cost of capital is only 6%. The regulated company will, therefore, be biased towards those (capital-intensive) technologies which increase (or 'pad') its rate base. In particular, its capital/labour ratio will be higher than that which minimises costs for the same level of output.

The A–J effect might be reflected in the following activities by regulated companies:

(*a*)   Resistance to peak-load pricing (which would reduce capacity entering the rate-base).

(*b*)   Willingness to maintain excessive stand-by capacity or provide excessive quality of service.

(*c*)   Resistance to capital-saving technology.

(*d*)   Reluctance to lease facilities from others.

(*e*)   Tendency to bargain less hard with suppliers.

(*f*)   Tendency to reach for additional business, if need be at rates below incremental cost, in order to extend the rate base.

How far are such phenomena observed in the US telecommunications system? The Bell System has long used peak-load pricing, in contrast to US electricity companies. Insofar as there is still spare capacity at night, there is scope for further offpeak reductions, but calculations do not suggest that an unconstrained profit-maximising company would practice much further peak-load pricing (Littlechild and Rousseau, 1975, Hopley, 1978).

AT & T executives are proud of the quality of service offered by their telephone system, but some outsiders would argue that it is too good (Kahn 1970/71, Vol. II, p. 53, Fn. 20, identifies Sheahan, 1951, as taking this view). To reduce congestion and avoid waiting lists costs money. Many private microwave systems, it is suggested, prefer to operate with a lower quality of voice transmission and a higher probability of congestion—This was indeed the basis of MCI's application to compete with ATT as discussed in Section 12.5. The fact that the New York Telephone Co. was unable to meet demand fully for a period during the early 1970s is not seen as counterevidence, but rather as an indication that the company failed to predict a sudden upturn in demand (at a time when its resouces were stretched anyway).

With respect to the use of capital-saving technology, the choice between satellites and submarine cables provides an opportunity to test the third aspect. It has been argued that

(*a*)   Satellites offer substantial cost savings which AT & T has resisted because they are less capital-intensive (Peck, 1970, but see Kahn, 1970/71, Vol. II, p. 51)

(*b*)   ATT favoured coaxial cable over less capital-intensive microwave radio relay during the first few years following World War 2, until other firms started to develop potentially competing networks (Shepherd, 1970, Scherer, 1970, p. 535)

(*c*)   In states with low allowed rates of return (hence high *net* cost of capital), AT&T tends to bury two cables to meet future demand, rather than adopt the cheaper alternative of reopening the trench and adding a second cable later (Scherer, 1970, p. 534)

(*d*)   In the choice between retransmitting satellite signals through established ground channels and TV stations and broadcasting direct to receiver sets, 'Comsat (the Communications Satellite Corporation) is predisposed for its domestic system toward the higher capital-cost intensive distributive system, for it would add more to its rate base than the direct broadcasting system' (Lessing, 1967).

On the other hand, some would argue that these claims are over-simplifications:

(*a*)   Engineering cost studies (DCF or PVAC) show that satellite is cheaper when a country has small amounts of traffic to many others (e.g. USA to Africa), since these can be handled by a single satellite instead of many cables. However, submarine cable is cheaper for a large volume of traffic between two countries. This is the case between USA and UK, where the FCC is opposing the unanimous view of Bell and the British Post Office.

(*b*)   Similar studies show that the relative costs of microwave and coaxial cable depend entirely on the nature of the terrain. Thus there is not a universal solution and most countries use either as appropriate. Radio relay is not necessarily less capital intensive. When relay stations are in remote locations, the cost of access roads, is many times that of the telecommunications equipment. Moreover, slowness to adopt new technology can be due to natural conservatism rather than an ulterior motive!

(*c*)   Adding a second cable later is not necessarily 'the cheaper alternative'. The cost of excavation is often many times the cost of the cable, particularly in urban areas. In this case, it is certainly cheaper to lay two cables than to reexcavate. This is why the British Post Office usually uses multiway ducts rather than direct burying; the first cable costs more to lay, but subsequent ones can be drawn into the duct at low cost without re-excavation.

(*d*)   Retransmitting signals through established channels involves the viewers in no extra cost and so is obviously sensible. The cost of *every* viewer needing to have a 'dish' aerial and microwave receiver is mind boggling. Some form of community distribution would be required, even if it is not the existing one!

The reluctance of common carriers to lease satellite channels from Comsat, rather than construct their own facilities, has been held to be a clear manifestation of the A–J effect (Kahn, 1970/71, Vol. II, p. 52, and the references therein).

It is frequently suggested that the Bell system operating companies pay more than they need to for equipment (despite studies showing higher quotations from outside suppliers, e.g. as quoted in Phillips, 1969, p. 671). If this is true, it is not only an attempt to increase the rate base, but also reflects the fact that the major supplier (Western Electric) is itself owned by AT&T. Since Western Electric is not subject to the same process of regulation, it is more convenient to earn profits at the level of manufacture (Westfield, 1965). However, others would argue that Western Electric can give lower prices and still earn high profits because of its vast scale of manufacture and steady stream of orders.

Do regulated telecommunications companies diversify into relatively unprofitable areas to expand the rate base and thereby increase prices in areas of monopoly power? Scherer (1970, p. 536) discusses one possible example:

> That AT&T behaves in this way was suggested by a study revealing that the telephone company during 1964 realised only a 3 per cent return on capital allocated to its TELPAK multi-channel business service and 2·9 per cent on its TWX (Teletype) service, both offered in competition with Western Union service and privately owned data transmission systems, while earning a 10 per cent return on its interstate telephone operations. Again, however, the evidence is ambiguous, for the rates of return are estimated on the basis of 'fully allocated costs' which include arbitrary overhead cost allocations. Since the TELPAK and TWX rates probably covered at least short-run marginal costs, it is impossible to reject the alternative hypothesis that AT&T's rate structure was nothing more than a straightforward discriminatory response to varying demand elasticities in different markets, demand being more elastic in the business data transmission field, where competition exists, than in long-distance telephony.

It can also be argued, of course, that fear of antitrust suits has prevented Bell from diversifying into areas which could be profitable, such as data-transmission terminal equipment.

Scherer's point is a valid comment on all the examples discussed in this Section. The illustrations we have given are broadly consistent with the predictions of Averch and Johnson. However, in view of the multitude of other possible explanations, they do not constitute overwhelming evidence of a serious bias in choice of technology.

13.4 Rate structures, cross-subsidisation and the burden test

During the 1960s, Western Union complained that AT&T was subsidising certain services offered in competition with Western Union by high profits in ATT's monopoly market of message toll telephone. In

1965 the FCC ordered AT&T to prepare a 7-way cost study, allocating all its costs between its even major services (Kahn, 1970/71, Vol. I, pp. 156–8). A comparison of these costs with the corresponding revenues was intended to show the relative profitability of each service. The rates of return so calculated varied from 0·3% on the bulk private line service TELPAK to 10·1% on wide-area telephone service WATS. The results also suggested that message toll telephone service was earning a return of 10%, while all other services taken together were earning less than 3%. These calculations appeared to support Western Union's contention that it was being subjected to unfair, subsidised competition.

The pros and cons of the 7-way cost study were much debated in testimony before the FCC and elsewhere (e.g. Froggatt, 1966, 1971, Baumol, 1966, 1971, Melody 1968, 1971, Bolter , 1978). Economists have been almost unanimous in criticising the attempt to allocate joint and common costs which the 7-way cost study represents. We have explained the reasons for this view in chapter 5. Briefly, such studies (i) refer to past expenditures (i.e. sunk costs) which are irrelevant to future decisions where prospective costs are required, (ii) do not reveal incremental costs, (iii) give a misleading calculation of average costs for a whole service and (iv) take no account of the conditions of demand.

The FCC did not finally decide whether the data produced by the 7-way cost study justified Western Union's conclusions. Kahn points out that these data need not do so; whether message toll telephone was subsidising other services depends on whether the latter covered their own *incremental* costs. Data on incremental costs were not provided at the time. However, in later testimony ATT described its long-run incremental cost (LRIC) analyses which were designed to answer such questions (Eastmond, 1970, Froggatt, 1971) and specifically to justify its claim of no cross-subsidisation (see Bolter, 1978, for an evaluation of FCC and other costing standards).

This issue may be restated as follows. Differences in profit rates on different services may well reflect price discrimination and 'charging what the traffic will bear', but this does not necessarily imply cross-subsidisation. A service is only being subsidised if it fails to cover its own separable costs.

These calculations may be related to a theoretical analysis provided by Baumol (1970), known as the *burden test*. Suppose that the question at issue is whether the provision (or continued provision) of a service constitutes a burden on the customers of other company services (i.e. whether it is being subsidised). The service is not a burden if it allows these other services to be provided at prices no higher than would otherwise be required—for example, if withdrawing the service in question would require the prices of all other services to be raised. To decide whether a specific service is or is not a burden it is necessary to calculate (*a*) the direct revenues it provides, (*b*) the indirect changes in revenues from other services resulting from cross-elasticities of demand, and (*c*) the corresponding incremental effects on costs of all other services and hence on the total revenue requirement of the firm.

Formally, an existing service *x* does *not* constitute a burden if

$$R_x > p_1 \, \Delta y_1 + p_2 \, \Delta y_2 + \ldots + p_n \Delta y_n - \Delta T$$

where R_x is the revenue from service x, $p_1, p_2, \ldots, p_n$ are the current prices of other company services, $\Delta y_1, \Delta y_2, \ldots, \Delta y_n$ are the changes in these services which would accompany the discontinuation of service *x* and ΔT is the effect on total revenue requirements of withdrawing service *x*.

Baumol emphasises that this test merely calculates the effect on customers of the firm's other services of continuing a particular service on specified terms; it does not by itself calculate the optimal terms on which to provide the service. This has led Braeutigam (1973 *a, b*) to propose a 'differential test' (as opposed to Baumol's '1-point' test) which compares provision of a particular service at a specific tariff u_x with the alternative of providing it at a different tariff t_x. This leads in turn to the search for a tariff u_x which satisfies the differential burden test for *all* possible alternative tariffs t_x. However, Braeutigam shows that, even if such a tariff can be found, it does not necessarily lead to economic efficiency, and in some circumstances may do exactly the reverse.

It should also be noted that the results of a burden test may depend on the precise definition of a service. For example, it may be easier for one component of private line service to pass the test than for total private line service itself. Because of the prevalence of common costs, the incremental cost of one component may be very low, whereas the incremental cost of all taken together may be high.

13.5 Changing FCC policy on competition and entry

Rapid and diverse developments in communications technology appear to have opened up new possibilities for competition. This has led to pressure on the FCC to allow the entry of new firms into the tele-communications industry to compete for markets traditionally supplied by AT&T. There is not space here to provide a complete history and evaluation of FCC policy, but the following brief account of four major decisions may be helpful. This summary is largely based on Kahn (1970/71, Vol. II, pp. 127–152); see also Trebing (1969). The account of the resale inquiry is taken from Criner (1977*a,b,c*).

(*a*)   The 'above-890' decisions (1959, 1960)

The development of microwave radio transmission made it technically possible for large users to set up their own facilities at lower cost than the tariffs charged by the common carriers. However, they needed FCC permission to use the radio spectrum. In its decisions of 1959 and 1960 the FCC finally granted frequencies above 890 MHz, on the grounds that adequate frequencies were available (contrary to the arguments of the common carriers), that private systems could supply service better tailored to individual requirements and that competition could provide a spur to manufacturing and new technology. The FCC refused to allow private users to share facilities to lower cost, thereby preventing additional competition in the common-carrier business itself. Nevertheless, to hold its large customers, AT&T responded with a new set of private line services at much more favourable rates, including TELPAK, and with wide-area telephone and data services (WATS and WADS). This price cutting led in turn to Western Union's accusations of cross-subsidisation as discussed in the previous Section. Section 13.8 of this chapter considers further the general problem of spectrum management.

(*b*)   The MCI cases (1969, 1970)

In 1967 a small firm Microwave Communication Inc. (MCI) applied to the FCC for permission to operate a public common-carrier link between Chicago and St. Louis. Proposed quality of service and back-up were minimal, and terminal equipment was not provided for subscribers; on the other hand, the proposed rates were half those of the existing carriers, interconnection with any subscriber equipment was allowed, and so too was the sharing of channels. The fact that MCI's proposed service was so very different from that of the common carriers led the FCC to give its approval in 1969, and, indeed, a year later virtually ordered the common carriers to provide interconnection facilities. The commission felt that competition would tap new markets, but it did not confront the key question of whether this competition would reduce quality and increase cost for the remaining customers of the common carriers.

Following the success of MCI's application, 33 other organisations filed for similar authorisation as specialised common carriers (SCCs).

At the time of writing, approximately six of these are still operating successfully (White, 1978). In addition, there are other SCCs whose primary interest is in transmitting video or television services.

The most spectacular application was the Datran proposal for a nationwide switched data network in direct competition with the Bell System. This did not prove successful. In 1976 Datran filed a petition of bankruptcy. Its major shareholder filed suit against AT&T, charging that AT&T had unlawfully kept Datran out of the market, and had used its monopoly power to impose predatory rates and block competition.

The battle is by no means over yet. Currently, MCI and AT&T are involved in court and commission hearings concerning MCI's attempt to introduce and expand its special service 'Execunet'. MCI is demanding that AT&T be required to interconnect MCI locally, while ATT is countering that MCI is attempting to open up to competition its message telephone service (MTS) and wide-area service (WATS), where AT&T has hitherto enjoyed a monopoly (White, 1978).

(c)   *Communications satellite (1965)*

Should satellites be operated by the existing common carriers or in competition with them? In 1962 Congress compromised by entrusting the international field to the Communications Satellite Corporation (Comsat), a new company to be owned equally by the existing carriers and the investing public. How far could Comsat compete with the carriers? The latter argued that Comsat should be a 'carrier's carrier', dealing only with them. In 1965 the FCC rejected this view, but nonetheless severely limited to exceptional circumstances the right of Comsat to deal directly with final users. In effect, this decision limited competition for the business of the largest users.

(d)   *Foreign attachments (1956, 1968)*

The Bell system had long refused to allow customers to attach their own equipment to Bell equipment, in contrast with the policies of the independent telephone companies. One such attachment was the Hush-a-Phone, a cuplike device which snapped on the instrument to provide an element of quiet and privacy. Over 100,000 Hush-a-Phones had been sold in the 30 years since their introduction. AT&T argued that good telephone service could not be provided if any kind of foreign attachments were permitted, and that such a device was not in public demand, or was unnecessary, or could be supplied by Bell if it were desired. In 1956 the FCC upheld the Bell position, but the Circuit Court of Appeals subsequently overturned its decision.

The Carterfone is a receiving and amplifying device making it possible to conduct 2-way conversations between telephones and private mobile radio systems. In 1968 the FCC held that ATT was not justified in refusing to allow its attachment to Bell equipment.

Later that year, the Chairman of AT&T professed his enthusiasm for the new era of interconnection between telephone company equipment and equipment supplied direct to the user by independent manufacturers. The Carterfone decision has led to rapid growth in sales of all sorts of attachments; this is known as the 'interconnect' business. Hasselwander (1978) has described the experience of Rochester Telephone Corporation in responding to this new environment.

(e)   *The resale inquiry (1976)*

Relaxations of policy in the late 1960s with respect to private network services (MCI) and customers' own terminal devices (Carterfone), did not extend to the use of leased facilities. Users could share lines on a not-for-profit basis (i.e. to reduce their own costs rather than to resell to the public), but middlemen were not allowed to set up such arrangements. Consortia of customers were not allowed to establish common networks under a 'facilities manager'.

This policy came increasingly under attack, mainly from user groups and telecommunications consultants wanting to take advantage of TELPAK bulk discount rates by setting up sharing consortia or reselling circuits. Wishing to defend the existing industry structure, AT&T argued that

(*a*)   The low price of TELPAK depended on not all channels being taken up (a low 'fill'), so that sharing or resale would increase the average 'fill' and necessitate price increases.

(*b*)   The shift of traffic to TELPAK and WATS would reduce AT&T's revenue, necessitating higher prices overall.

(*c*)   Third parties would disrupt the dialogue between carrier and ultimate user, to the detriment of long-term planning.

Against these points, it was noted (i) that a revised higher-price TELPAK was necessary in any case, (ii) that the claimed loss of revenue was only 0·76% of total Bell System annual revenue and (iii) that some brokers might become specialists with a better knowledge of future demand. The Office of Telecommunications Policy (a government department, now replaced) advocated a policy of unrestricted resale, arguing (i) that sharing and resale would lead to more efficient use of telecommunications plant, (ii) that brokerage would make low-cost transmission available to a greater proportion of users, eliminating bulk discounts which favour larger users only and (iii) that new marketing and management expertise would increase aggregate demand for telecommunications.

In 1976 the FCC decided that sharing on a not-for-profit basis could be extended from private lines to bulk discount lines, and gave explicit approval to the establishment of nonprofit intermediaries to co-ordinate the sharing arrangement. Only common carriers were allowed to resell for profit to the public, but the FCC here reaffirmed its policy of open entry: applicants would be authorised to act as common carriers without the need to show that the proposed service was unique, or that an unmet need existed or that existing carriers would not be adversely affected. However, the FCC did not call for removal of resale prohibitions or the long-distance message telephone service, which remains a monopoly of the established telephone companies.

13.6 Natural monopoly

A natural monopoly is usually defined as an industry in which economies of scale are so great, compared to the size of the market, that it is inefficient to have more than one firm producing the industry's output, and, in fact, only one firm would be able to survive in such an industry. Communications is generally held to be such a natural monopoly, yet nowadays in the USA we find competitors successfully challenging the telephone company and customers choosing to serve themselves. How, if at all, can this paradox be explained? What are the implications for policy? Kahn suggests several possibilities:

(*a*)   The new entrants might be mistaken as to the true levels of cost and demand. Undoubtedly some are over–optimistic, but too many others have been successful for this explanation to be generally acceptable.

(*b*)   Incumbent telephone companies might be charging such exhorbitant prices that even higher-cost entrants can make a profit. If so, a reduction in profit levels would suffice to put the newcomers out of business.

(*c*)   The new entrants might have more advanced (lower-cost) technology which yield lower prices only because the Bell system's traditional depreciation policy has not been adequate to reflect this unexpected technological progress, leaving high book costs to be covered. Here, a revised bookkeeping approach would suffice to make the Bell system competitive.

(*d*)   AT&T's policy is to charge a uniform price per mile nationwide, despite significant differences in cost (for example, depending on volume of traffic). If new entrants confine themselves to the heavy-traffic routes, they will have lower average costs overall (this practice is known as 'cream skimming'). Here, the remedy for AT&T is to price differentially, relating price more closely to cost for each service.

(*e*)   The new entrants could be providing different services from those met by the established companies. This seems to be true of MCI and the suppliers of auxiliary equipment. These new areas are thus not necessarily within the scope of the natural monopoly. Competition may be appropriate in the new areas, although the traditional services of the established companies remain a natural monopoly.

(*f*)   The provision of inferior services may be commercially acceptable only if Bell stands ready to provide emergency back-up service, but, providing Bell charges a price above incremental cost for this service, it does not affect the argument of the previous paragraph (and otherwise Bell should adjust its prices accordingly).

(*g*)   The new entrants may be providing the same service as the existing companies and operating at approximately the same average cost. With the new technologies and the growth of traffic, economics of scale may be exhausted at a smaller proportion of a larger market, thereby making room for several companies to compete. Waverman (1975) has argued that this is true for microwave transmission. In this case, the area of natural monopoly is nowadays more restricted than formerly. It still seems to encompass local exchange service, and perhaps transmission service on low-density routes, but not most other aspects of service.

These various factors explain how competition can be consistent with natural monopoly in at least part of the communications system. However, the actual extent of natural monopoly is by no means clear, and in Kahn's view the experience of the last quarter-century demonstrates the virtue of free entry and competition as a device for innovation. The likely existence of natural monopoly in part of the communications system is not an argument for retaining monopoly control throughout the system by prohibiting entry. Indeed, there is a case for allowing competition across-the-board precisely in order to establish the limits of natural monopoly.

13.7 Sustainability of natural monopoly

Not surprisingly, AT&T has not welcomed the prospect of extensive competition. In recent years, several economic theorists, many associated   with Bell Telephone Laboratories, have extensively analysed the meaning of natural monopoly and its relationship to new entry. We shall briefly survey their results to date.

The definition of a natural monopoly given at the beginning of the previous Section seems to imply three propositions:

(*a*)   In a natural monopoly, it is cheaper for one firm to produce any specified output than for two or more firms to do so.

(*b*)   A natural monopoly is characterised by economies of scale.

(*c*)   A natural monopolist can charge prices which cover his total costs and still make entry unattractive to any rival firm.

It used to be thought that these three propositions are equivalent, but they are not.

A cost function such that a single firm can produce at lower cost than two or more firms is said to be *subadditive*. Baumol (1977) has shown that economies of scale are neither necessary nor sufficient for subadditivity, hence are neither necessary nor sufficient for monopoly to be the least costly form of productive organisation (see also Sharkey and Telser, 1977). As we have seen, econometric studies suggest that AT&T and Bell Canada do have economies of scale in their system taken as a whole, though perhaps rather slight, and in certain sectors such as transmission these economies are held to be exhausted at relatively low levels of output. Thus, for present purposes, the significance of Baumol's result is that a monopoly may be the most efficient way to supply telecommunications, even though this monopoly does not exhibit economies of scale.

Neither is it true that a natural monopolist with a subadditive cost function can ward off potential competitors. A set of prices for its various services is said to be *sustainable* for an established firm if and only if (i) it makes a profit at those prices and (ii) no other firm, having access to the same cost function, can offer lower prices, capture part of the business, and still make a profit. Panzar and Willig (1977) show that, even if the cost function is subadditive, the established firm may not be able to choose a sustainable set of prices.

A basic dilemma thus confronts the regulator of a natural monopoly. Free entry may encourage cost control and stimulate innovation. Yet, free entry may permit firms with neither new products nor improved technologies to enter the industry, and thus cause higher industry costs, higher prices, and net welfare losses.

The clear implication is that restrictions on entry may be in the social interest, even though new entrants could make a profit. On the other hand the dynamic advantages of innovation may outweigh these static efficiency considerations, but the static model cannot evaluate them, as the authors recognise. Further, to rephrase Kahn's observation somewhat differently, one of the functions of competition is precisely to establish whether or how far an industry is in fact characterised by a subadditive cost function; it will not do to take the established firm's word for it.

13.8 Management of the radio spectrum

The rapid growth in all aspects of telecommunications has led to increasing pressure on the limited availability of radio spectrum (i.e. the full range of radio waves that may theoretically be used to transmit information by electromagnetic energy). Existing users (including police and defence services) would like to expand, but cannot; new users (such as mobile-radio services) cannot obtain more than a small foothold.

This problem has become particularly acute in the USA, where it has led to much discussion about alternative methods for utilising and allocating radio spectrum. In particular, economists have been urging that radio frequencies be allocated by a market system, instead of by a government agency. It would appear that this and related proposals for change are now being studied seriously by the US Government. Other countries are likely to have to face the same problem in the near future.

The following account is based on a paper by former FCC Commissioner N. Johnson (1969). This gives an excellent and comprehensive statement of the difficulties of the present system for radio-spectrum management in the USA, and outlines various proposals for reform. Those economists arguing for a market system include Coase (1959, 1962, 1966), Levin (1962, 1966, 1968, 1970) and L. Johnson (1967) (see also de Vany *et al.* (1969) and the report of the President's Task Force (1968, Chap. 8)).

The Communications Act of 1934 charged the FCC with the efficient allocation of the radio spectrum among competing claimants. The act was further designed to encourage the conservation of valuable frequencies, while promoting the efficient development and utilisation of the spectrum.

Under the present system, potential users of the spectrum must apply to the FCC. A choice must be made between users, either on the basis of 'first-come first-served', or else on the basis of 'superior qualifications'. In practice, according to Johnson, the FCC has developed no consistent principles of licensee qualifications, so that an irrational and chaotic system has resulted. This, in turn, has the following harmful consequences:

(*a*)   Inefficiencies and distortions in the economic system as a result of the arbitrary allocation of frequencies between users.

(*b*)   Complete absence of economic considerations in evaluating 'at the margin' a unit of spectrum in alternative possible uses.

(*c*)   No incentive for present users to economise on spectrum (e.g. by use of transportation instead), and positive incentives to 'stockpile' spectrum for the future.

(*d*)   No systematic provision for transferring spectrum from present users to new or alternative users, nor indeed any method for evaluating benefits flowing from any particular (actual or potential) allocation of spectrum.

(*e*)   No possibility of private exchange of reallocation of spectrum by incumbent users.

(*f*)   No incentive for more intensive use of spectrum through sharing, modifying acceptable level of interference, redesign of systems, etc.

(*g*)   No capacity for FCC to evaluate proper mixture of newer (high-frequency) parts of spectrum.

(*h*)   Inadequate preparation for possible explosion in future demand e.g. 100 or 1000-fold increase in mobile communications (since Johnson's paper, such an explosion has, in fact, begun to take place. The number of citizens' band licences in the US increased more than 10-fold in the 3 years 1974–77. See 'Will Britain join the Citizens' band?', 1977).

Because the spectrum is inadequate to accommodate all who would like to use it, choices must be made as to types of use and individual users. In other parts of the economy, such scarce resources are allocated between claimants by means of a market system. A price is established which is related to the benefits conferred by use of the resource. Alternative users always have the ability to bid away resources by paying a higher price. Such a price system is absent for the radio-spectrum resource.

A group of economists led by Professor Coase has argued that the radio spectrum should be sold by government to private users. Property rights in the spectrum would have to be defined, and the law would have to develop a system of enforcing these property rights. But after the initial sale from the government, transactions between users would be out of the control of a spectrum manager. Transactions of all sorts, sales, sharing, splitting, trades and so forth, would be carried on with traditional institutions: the legal protection of property and the allocation of resources by the price system.

Johnson points out an analogy with the distribution of land in the western part of the USA at the end of the last century. Initially, the government delineated property lines, but ownership rights were then allowed to determine future land development. The merits of the market allocation proposal are fourfold:

(*a*)   Elimination of the present expensive and time-consuming allocation procedure.
(*b*)   Greater efficiency in utilisation of spectrum (by individual users and as a result of transfers between users).
(*c*)   Avoidance of any threat to freedom of the speech inherent in a system of government approval.
(*d*)   Avoidance of arbitrary enrichment of certain users through free disposition of valuable frequencies.

Various problems are sometimes thought to remain, pertaining (i) to the engineering difficulties of creating property rights in radio spectrum, (ii) doubts as to whether the market functions efficiently, (iii) doubts as to whether ultimate consumers of TV and radio programmes can effectively make known their preferences and (iv) social as opposed to profit considerations.

There are also political objections from those with vested interests in the present system and those who believe the spectrum is too precious to be allocated by the market.

Alternative means of improving the present system have been proposed, including (i) the rental (rather than outright sale) of frequencies, (ii) the use of 'shadow prices' to stimulate the market mechanism (by estimating the value of frequencies to different users), (iii) the creation of a new government department with the sole duty of managing the spectrum, (iv) the development of improved and consistent guidelines for allocation of frequencies by the FCC and (v) the introduction of user fees more closely related to market values.

Johnson's own preference is for a reform of the existing FCC procedures. He believes the Communications Act envisaged that spectrum users, in exchange for frequencies, would operate in the public interest rather than in their own private profit-maximising interest. This does not, however, require *free* access to the spectrum, so that higher user fees and shadow pricing have an important role to play. Most important, however, is that the FCC rethink its concept of the 'public interest' and its implementation in spectrum management.

In the decade since Johnson's paper was published, the FCC has clearly had to rethink its position on many issues, but the central problem of spectrum utilisation and allocation still remains to be tackled.

Some alternative views

14.1 Introduction

The first part of this book introduced the basic tools of telecommunications economics—the demand functions, production functions and cost functions (chapters 3–6). It was then shown how these tools could be used to characterise pricing and investment policies to maximise proft (chapters 7 and 8). Next, the ideas of welfare economics were expounded, showing what pricing and investment policies would be necessary to achieve a social optimum (chapters 9–12). Finally, the experience of regulating the privately owned US telephone industry was examined (chapter 13).

Most of the ideas discussed so far are fairly well established in economics, although their application to telecommunications is very recent; almost all the empirical work referred to has been done in the last decade. The recommendations for socially efficient or optimal pricing derive from the body of welfare economics which has been developed over the past half century. Welfare economics, in turn, is firmly set within the so-called *neoclassical approach* which has characterised economic theory over the last century.

The basic assumptions of neoclassical economics are (i) that the consumers and firms in the economy attempt to *maximise* their utility or profit by efficiently allocating the scarce resources available to them, (ii) that they make these choices in the light of *full (or adequate) information* about the alternatives available and (iii) that the interrelationships between these decision makers can best be understood by the technique of *static equilibrium* in the industry or in the economy as a whole. Thus it is that almost the entire body of knowledge in telecommunications economics, like the discipline of economics itself, is based on these three key concepts of maximisation, full information and static equilibrium.

In no scientific discipline do ideas remain constant, however, and economics is no exception. An increasing number of economists began to feel that the traditional neoclassical framework was too restrictive. It failed to capture certain economic phenonmena which seemed to be important. This feeling was reflected in several attempts to extend, and in some respects to replace, the three key concepts identified above. In this book, we shall not be concerned with the first and most important of those attempts, namely by John Maynard Keynes, which resulted in the development of a quite separate theory of *macroeconomics* to deal with aggregates such as labour, capital, unemployment and money.

Our interest here lies in the ideas, independent and in part unrelated, pioneered by a dozen or so other economists over the last 15 years. Their work may conveniently be divided into three categories, each of which will need to be further subdivided.

First, there is the attempt *to extend the concept of maximisation* from consumers and profit-maximising firms to explain also the behaviour of nationalised and regulated industries, charities, local authorities, politicians, civil servants and governments. Within this category three subcategories may be identified:

(*a*)   The role of *property rights* as affecting the incentives, and hence the behaviour, of members of these organisations (Alchian, 1965, Alchian and Kessel, 1962, Alchian and Allen, 1974).

(*b*)   The theory of *public choice* in the political sphere to complement the theory of private choice in the market (Buchanan, 1968, Buchanan and Tullock, 1962, Buchanan and Tollison, 1972).

(*c*)   The theory of *economic regulation* to explain how and why governments actually *do* behave, as opposed to a normative (welfare economic) theory of how they ought to behave (Stigler and Friedland, 1962, Stigler, 1971).

Second, there is the attempt *to abolish the assumption of perfect information*, and to rebuild economics on the assumption that its actors are beset by uncertainty and subject to error. Within this category are at least four subcategories:

(*a*)   The theory of *optimal search for information*, which explains why the owners of resources (including labour) may choose to keep them unemployed in order to wait for a better offer (Stigler, 1961, Alchian, 1969).

(*b*)   The possibility and nature of *general equilibrium under uncertainty*, which is possibly only of a temporary nature (Hahn, 1973, Grandmont, 1977, also Hayek, 1937).

(*c*)   The implications of uncertainty for the *nature of decision making*, Shackle, 1961, 1972).

(*d*)   The interpretation of *cost as subjective* and its implications for monitoring the behaviour of subordinates (Thirlby, 1960, Wiseman, 1957, Buchanan, 1969, Buchanan and Thirlby, 1973).

The third category of work also stems from removing the assumption of perfect information, but its consequence is of significance in its own right, namely, *to complement the theory of static equilibrium by a theory of dynamic process over time*. This work is associated with the 'Austrian' school of economists (Menger, 1871, Mises, 1949, Hayek, 1948, Kirzner, 1973; see also Littlechild, 1978). According to their view, the knowledge in society is necessarily dispersed: different men know different things. Mistakes are made through lack of knowledge, but men also learn from their mistakes. In fact, competition may be seen as a process of discovering which techniques are best which products customers require etc. This process necessarily takes place over time, and the conditions of equilibrium (if ever attained) would represent the culmination of this process. It is, however, the process rather than the equilibrium which is of major interest.

These new developments in economics are bound to be reflected, sooner or later, in the economics of telecommunications. The purpose of this final chapter is to appraise some of the likely developments. We are led to ask different questions about the past and present behaviour of telephone administrations and government policy towards the telecommunications industry. Naturally, the answers show up this behaviour in a somewhat different light from that in which it has hitherto appeared. Moreover, we are led to a somewhat different interpretation of socially optimal behaviour. Whereas welfare economics placed greatest emphasis on characterising optimal pricing and investment policy of telecommunications administrations, these new ideas place greater emphasis on the design of the institutional framework within which decisions about telecommunications are made.

14.2 Property rights and behaviour

When deciding which course of action to choose among the alternatives available, a person will take into especial account the consequences for himself. If he bears only *some* of the consequences of his action, he will in general act differently than if he bears *all* the consequences. The system of property rights determines which consequences he has to bear. Accordingly, changes in the system of property rights are likely to change behaviour.

One instance of this proposition has already been discussed, namely, the Averch–Johnson effect. If a company is no longer entitled to all the profits which accrue from its actions, but only profits up to a fixed rate of return on some base, then it will adjust its activities to increase this base, for example, by choosing capital-intensive methods of production. It will also have less incentive to discover ways of reducing costs, or increasing profits, if the savings have to be returned to the customers. Alternatively, it will be inclined to 'mop up' such savings by expenditures which generate pleasure for its members, e.g. expense-account meals and travel, or pleasant office facilities.

Such changes in behaviour are even more likely if the company is not regulated, but is nationalised. Here there is no owner to claim the residual, or, more precisely, the voter is less effective an owner than the shareholder. With reduced profit incentives to monitor the behaviour of the managers, there is more opportunity for the managers to act in ways which are congenial to themselves, rather than in ways which maximise profits and minimise costs. This is not to say that managers or employees of nationalised industries or municipal organisations are different or less worthy kinds of people from those who work in regulated or unregulated private industries. Rather, the proposition states that the same people, faced with different incentives and different rewards, will act differently. In any situation they will act to maximise their own utility; if a change in property rights changes the utility of any action then it is likely to change behaviour.

There has been a great deal of empirical work to test this hypothesis, especially in the US electric-power industry (Peltzman, 1971, De Alessi, 1974). Broadly speaking, the hypothesis is supported as De Alessi indicates:

> The main findings pertaining to the electric power industry are as follows. The regulation of privately-owned firms seems to yield, among other things, slightly lower structure of rates which is more favourable to the larger users, and to industrial users in particular, relative to other, more numerous user groups. There is also growing support for the Averch–Johnson over-capitalization hypothesis. The information regarding the consequences of government ownership is richer and more varied, particularly as a result of Peltzman's imaginative research. More specifically, the evidence suggests that municipal firms, relative to privately-owned regulated firms, in general will: charge lower prices; have greater capacity; spend more on plant construction; have higher operating costs; engage in less wealth-maximising price discrimination, including fewer peak-related tariffs; relate price discrimination less closely to the demand and supply conditions applicable to each group of users; favour business relative to residential users; offer a

smaller variety of output; change prices less frequently and in response to larger changes in economic determinants; adopt cost-reducing innovations less readily; maintain managers in office longer; exhibit greater variation in rates of return.

In general, the direct evidence supporting any one of the individual conclusions noted above is not overwhelming. Levels of significance often are not low enough, while the variables used and their measurement frequently lack sufficient precision and theoretical justification. Nevertheless, the extent to which all the various bits of evidence support each other and fall into a systematic, predictable pattern is impressive.

Many telecommunications engineers appear to believe that the type of ownership is unimportant, providing there is 'unity of control' and the engineers are allowed to get on with the job. Thus Robertson (1947, p. 48) writing about Britain:

> Again and again in the early history of telecommunications in this country this dichotomy is revealed. Policy direction was short-sighted, acquisitive, blundering and ignorant; technical and engineering work was based on sound, scientific principles, and was far-sighted and highly intelligent. That was the root of the trouble for years; the engineers knew their job, and the others did not.

Many businessmen, on the other hand, feel that the present US telephone system is superior to foreign systems precisely because it is privately owned (Reuss, 1975). The early account of telephone systems by Kingsbury (1915) contains a chapter 'The telephone and governments' which is sympathetic to this view, and remarkably prescient of the property-rights hypothesis. He suggests that

> ... with all the varying results and divergent conditions there would seem to be a consensus of opinion, confirmed by statistics and experience, that [telephone] development is greater and service better under private enterprise than under Government management. [p. 510]

One reason put forward is that 'the telephone exchange service requires of the public an active participation'. The subscriber is painfully conscious of any delay in providing a connection, and if the system is government owned he naturally complains to the government, but this is not always helpful.

> The critic generally has his own remedy to recommend; his complaints find expression in Parliament or Press, and under the pressure of public opinion, a Government department may be called upon to carry out some measure which experts are aware may not conduce to real improvement. In such ways the parliamentary interference, which is generally regarded as a safeguard, may be a cause of increased difficulty.

The fact that the service is undertaken without any incentive of gain leads the subscriber to infer that his interests are not being studied. Appeals to the self-interest of a private supplier are expected to be productive of some good, whilst from a public body reasonable attention to complaints is regarded as hopeless. The very general assumption that Government officials habitually disregard complaints or are inattentive to public demands is by no means correct—so far as Great Britain is concerned at any rate. Brault, writing of the German telephone system in 1890, says that the establishment of the lines had been rapid but the material was inferior and the service defective. He expressed the opinion that the autocratic methods of Government officials lacked the adaptability which was necessary to meet the requirements of subscribers. The organisation of a Government department is necessarily less flexible than that of a private company. The will may be there, but the way is more difficult. The inertia is greater, the start less prompt, and the accomplishment slower and less certain. Criticism of Government service is more ample, less judicial, and far less effective than that of private enterprises.

Kingsbury goes on to suggest that the efforts by government telephone administrations to install automatic switchboards is partly accounted for by the inefficiency and cost of government employees.

This inefficiency is not of the individual, but results from the absence of the discipline, control and incentive which are available in private enterprise. [p. 512]

These observations are remarkably in sympathy with current thinking on property rights by economic theorists.

14.3 Empirical evidence on ownership and performance in telephone systems

There seems to have been no systematic empirical work to test whether the property-rights hypothesis is as valid in telecommunications as it appears to be in electric power. The major difficulty is that nowadays there is typically only one form of ownership within any country. Only a handful of countries (including Canada, Denmark and Portugal) have privately owned telephone systems working alongside government-owned systems. To obtain any estimate of the effect of ownership it is necessary to make international comparisons. This raises other difficulties, for any observed differences in behaviour may be due to differences in other economic and social conditions between the various countries. The effects of such different conditions needs to be removed mention briefly some interim results of work in progress by the present author.

The Swedish Telecommunications Organisation (Roos *et al.*, 1976) has produced a survey of telephone tariffs in 55 countires, as they obtained on 1 September 1975. These data must be treated with some caution, since they are based on a postal questionnaire which elicited only a 50% response rate, and presumably no further checks were carried out into the accuracy and comparability of the information provided. Nevertheless, these data are very helpful and will serve for a first analysis.

A comparison of the tariffs adopted by privately owned telephone systems with those of government-owned systems reveals that a significantly higher proportion of private systems:

(*a*)   Differentiate the subscription rate with respect to category of subscriber.
(*b*)   Offer a reduction for a party line.
(*c*)   Differentiate call charges by time of day.
(*d*)   Employ more than two time-of-day rate periods for calls.

However, it is also true that differences in tariff policy are associated with differences in national income, with higher-income countries being more likely to behave in the above ways. If the analysis is restricted to those countries in the higher-income half of the sample available, type of ownership no longer significantly affects behaviour of types (*a*) to (*c*). On the other hand, private systems are still much more likely to employ more than two time-of-day rate periods.

There have been several studies relating telephone penetration to gross national income, as discussed in chapter 2. Littlechild (1976) has shown that after taking account of differences in national income, elasticities of telephone penetration with respect to population density and with respect to income are significantly higher for private systems than for government-owned systems. This suggests that private systems are more sensitive to differences in cost and demand. Whether telephone penetration is higher in a private system depends on the levels of the explanatory variables; rough calculations suggested that under British conditions of 1973 the average government system would exhibit a penetration of 21 telephones per hundred population, whereas the average private system would exhibit a penetration of twice that level. Penetration in Britain in 1973 was 34 telephones per hundred population. Thus, to the extent that any faith can be placed in these calculations, they suggest that the Post Office has achieved higher telephone penetration than the average government telephone system, but that a privately owned system might have achieved an even higher level.

These results are consistent with the hypothesis that the difference in property rights between government and private system affects behaviour and performance in definite and predictable ways. They do not, of course, prove that a certain type of ownership is best, but they provide a relevant contribution to the discussion of that question.

It is worth pointing out that for many years the British Post Office was starved of capital by the government, despite long waiting lists for connections. The Bell System, on the other hand, was able to raise investment, install plant and sell telephones whenever the prospect of profit warranted it. Thus, whether the level of telephone penetration in Britain is best explained by the *internal organisation* of the Post Office or by external *constraints* on the Post Office is not yet established. The empirical results pertain only to the system as a whole.

14.4 Property rights and externalities

It was shown earlier that subscribers making telephone calls during periods of congestion did not have any incentive to take into account the additional costs of congestion imposed on other subscribers. Subscribers have no right (as the law presently stands) to prevent calls being made, but they do, in principle, have the right to negotiate with other subscribers and to offer them financial inducements to limit their calling. If this were practicable, it would mean that the 'social' consequences of making a call would indeed be brought to bear on the caller, in the sense that he would have to forego the financial inducement offered, which in turn, would reflect the congestion costs imposed on others.

Of course, such a system is not practicable because there are so many subscribers that negotiation would be uneconomic. In other words, the existence of *transactions costs* precludes the market from achieving a socially efficient outcome. It is for this reason that the telephone administration is urged to adopt congestion pricing, as a substitute for the ineffective market.

There is, however, an alternative possibility of handling externalities within the market mechanism. If subscribers incur severe congestion costs, the value to them of membership in the telephone network is that much less. The maximum amount they are willing to pay as a subscription fee is reduced. Suppose that the telephone network is privately owned, and that there are no constraints on the owner maximising profits. The existence of congestion means that the owner's profits are reduced. In other words, the owner has an incentive to set prices and/or limit the volume of calls in relation to capacity so that the value of the network to the subscribers is as high as possible, because his own profits depend directly upon this value.

Thus we see that the regulation of profits, or constraints on the pricing system adopted, may have the effect of reducing the financial incentive to provide a telephone system free of congestion. If the system is nationalised, this financial constraint is almost eliminated (almost, not entirely, because the administration may urgently need money to finance expenditure elsewhere, but this incentive is presumably less than that of profit to an owner).

Once again, this analysis is not, in itself, an argument for a privately owned system. There are obvious disadvantages of private ownership insofar as a local telephone network enjoys a natural monopoly. The analysis is intended to show how the market can, in fact, handle externalities in ways which are not at first sight obvious. It also suggests an explanation of why grade of service is so poor in many developing countries, where the telephone systems are owned and operated by the government. A testable prediction of this theory is that grade of service is worse in government telephone systems than in private ones. (In testing this prediction, one would need to take into account the possibility that poor forecasting could inadvertently provide a good grade of service, and that improved forecasting can allow operation with a nominally worse grade of service.)

The analysis of this Section is equally applicable to (positive) externalities associated with membership in the network. That is, a private system would have a financial incentive to introduce new subscribers who increase the value of service to others, providing it could 'cream off' this additional value, either by higher subscription rates to existing subscribers or in some other way.

The ideas of this section were originally applied to road congestion by Knight (1924) and have been greatly developed by Alchian (1965) and Alchian and Allen (1974).

14.5 Economic regulation and competition

The purpose of regulation is presumably to keep down prices and profits to a.lower level than they otherwise would be. Stigler and Friedland (1962) disconcerted many proponents of regulation when they compared regulated and unregulated electric utilities in the USA in the period up to 1937, and found no evidence of any significant difference in the average level of rates, in the structure of rates or in shareholders' profits, between the two groups of companies. Regulation appeared to have no effect.

Stigler (1971) was later led to a startling explanation: 'as a rule, regulation is acquired by the industry and is designed and operated primarily for its benefit'. Government regulation can legalise and facilitate collusion to increase profits and it can prevent the entry of new competitors. On the other hand, it involves some loss of control to a government agency, and it has to be paid for in some way, typically by carrying out activities which generate benefits for those with power to bestow regulation on the industry. This idea is sometimes known as the 'capture' theory of regulation, it being suggested that the regulatory agency is captured by the industry which it regulates.

The thesis was extended by Posner (1971) to suggest that *any* special-interest group could achieve regulation, providing it were sufficiently well organised. The general thesis is that the intervention of government in the market does not represent a consensus by a society as a whole that a certain kind of intervention will generate benefits for all; rather, it reflects the success of one or more particular political pressure groups in obtaining favourable legislation at the expense of other, less powerful, groups.

This leads us naturally into an application of the Austrian view of competition as an active process of betterment taking place over time. Producers ceaselessly compete to offer the best terms to consumers by cutting prices, modifying the product, discovering new products, encouraging the customer to develop new tastes etc. Firms would presumably prefer *not* to have to compete in this way, but, unless they can prevent new competitors from taking their business, they have no option but to fight to remain one step ahead. One form of relief, however, is to collude with other existing producers, jointly attempting to limit competition among themselves and to hinder the entry of newcomers. Typically this is ineffective, at least over a long period, and may be illegal, unless some form of government approval and support can be obtained. A grant of monopoly power is therefore sought, perhaps in the form of regulation.

This view of competition and the role of government is quite different from that envisaged in welfare economics. There, perfect competition consisted of a large number of firms producing an identical product with access to identical technology and with perfect information about consumers' tastes and the behaviour of each other. This resulted in single price equal to marginal cost, sufficient to yield a 'normal' profit but no more. In this alternative view of competition, competitors may differ considerably in their knowledge of technology and consumer tastes and in the kinds of products they offer. They have to discover what it is that consumers want, and they have to meet these needs before their competitors do, or on better terms. Those who are successful make high profits, constrained only by the prices which the next-best competitor charges; those who are unsuccessful lose money and eventually go out of business. Since tastes and technology are always changing, this competitive process never ends.

Welfare economics predicts that if economies of scale are sufficiently large a natural monopolist will emerge charging monopoly prices; to prevent this outcome, regulation is needed. 'Regulation is the law's substitute for competition' is apparently the legend on the wall of the Michigan Public Service Commission's hearing room. The alternative view put forward here does not concede that natural monopoly means the end of competition, for without legal restrictions there is still scope for rivalry to become or replace the natural monopolist, and the ever-present threat of changes in tastes and technology make this position no sinecure. Regulation, in this view, is sought precisely to *avoid* the rigours of competition.

14.6 Competition and regulation in US telecommunications

Which of these alternative theories is more applicable to telecommunications? Gabel (1969) is in no doubt as to the answer to this question:

> The available historical evidence suggests that, at least in the communications industry, regulation has served to stabilise price and earnings of the carriers, has inhibited innovation in rate structures, and has protected the carriers from the competitive inroads of private manufacturers and suppliers.

It will be enlightening to summarise the argument of Gabel's paper.

Current discussions of telecommunications policy often assume that competition proved not viable. For example, the Director of the Office of Telecommunications Management stated (President's Task Force, 1968):

> Experience going back some seventy years has demonstrated that competition in the provision of local telephone service was inherently inefficient and led to poorer quality service at higher cost.

In fact, as Gabel shows, the expiry of the basic Bell patents in 1893–4 led to an extremely rapid growth in number of telephone stations, in both existing and new markets; severe reductions in rates, in exchanges with competition and also in those without; and a new and substantial investment by the Bell System in fundamental research. Although Bell argued that plant was wastefully duplicated, the independent companies claimed that the extent of duplication was minor, and that all facilities were efficient, necessary and provided at reasonable cost.

The Bell system initially (1893–1907) met this competition by a cut in rates and a rapid expansion in service. It also engaged in an active propaganda campaign against the independents, refused to connect with the independents for long-distance service, and refused to sell telephone equipment outside the system. In the later competitive era (1907–20), after the (re-) appointment of Theodore Vail as President, the rate of expansion was slower, and the Bell system launched an aggresive policy of acquiring independent telephone companies. The most significant reversal of policy, however, was the espousal of government regulation. The Interstate Commerce Commission (ICC) was attempting, with some success, to stabilise markets and price structures and to avoid 'price wars' in the railroad business. Vail early saw the possibilities of effecting such normalisation and stability in the telephone industry. In 1910 the ICC was given authority over interstate telephone companies. Between 1910 and 1920 31 states established authority for regulating interstate operations of telephone companies—'Business courts', wrote Vail, 'will soon bring order and security out of the present uncertainty and be a bulwark against future economic disturbance'. (AT&T Annual Report, 1914, pp. 47–9, quoted by Gabel).

Gabel summarises as follows:

In a sense all business enterprise is a flight from competition. The penalties of competition—low or nonexistent profits—may be avoided by superior efficiency, by product innovation or differentiation, or by attenuation of the competitive process through control over supply and price wielded monopolistically or through conspiracy or tacit understanding with competitors. Confronted by the vigorous competitive inroads of independent operating companies, the Bell System sought to escape the unaccustomed hardships of competition by acquiring competitors, by limiting their markets and their services, and by espousing the development of governmental regulatory functions. The public service commissions, which ultimately stabilized rates and earnings, adopted the norms of business policy urged by the System and imposed strictures on the 'unintelligent competition'. The advantages thus gained by the Bell System over its remaining competition have been parlayed into a practically unassailable market position fortified by political and legal ramparts.

The thesis has been posed that telephone competition during the years 1893–1920 was neither inefficient nor costly but was, on the contrary, productive of benefits sharply outweighing its costs. It was not just the working out of the competitive market process toward the emergence of inevitable 'natural' monopoly which destroyed the structure that permitted competition to flourish and its benefits to be enjoyed; it was as much a poorly conceived, Bell-inspired, protectionist regulatory policy which failed to preserve such competition as might usefully have served the public interest.

14.7 The role of the British Post Office

If further proof were needed that government departments scarcely fill the role advocated for them by welfare economics, a cursory examination of the history of the British Post Office would suffice. The following account is largely based on Robertson (1947), who in turn acknowledges works by Kingsbury (1915), Baldwin (1925), Murray (1927, Whyte (1930), Wolmer (1932) and Crutchley (1938).

The telephone was initially regarded by the Post Office with some scepticism. This view soon changed when it was realised that the telephone was likely to imperil the substantial revenues derived from the telegraph, which was a Crown monopoly by the Telegraph Act of 1869 (this Act, in turn, had been passed to protect the Crown revenues deriving from its postal monopoly). The Post Office took the telephone company to Court and in 1880 obtained judgment that, within the meaning of this Act, the telephone was in fact a telegraph, a telephone conversation was in fact a telegram, and the telephone company and its subscribers were, therefore, infringing the Postmaster General's monopoly.

The Post Office was faced with a dilemma. To throttle the telephone system completely was politically impracticable, but to operate the telephone system itself demanded a degree of prevision running directly counter to the long-established caution of government departments in the nineteenth century. It was decided to license private telephone companies for a limited period, but without granting rights to erect poles or wires on other peoples property, and the construction of trunk lines was forbidden.

The limitations on trunk lines could not be maintained for long. In 1892 the Postmaster-General announced that the government proposed to take over the trunk telephone lines; his reason was quite explicit:

> .. that there is a real danger of the valuable public property – the telegraph system – being injured by the extension of this telephone system, is proved by the fact that wherever the telephone system has been principally developed, there the growth of the telegraph revenue has been checked . . . If the telephone companies were in communication with all the large towns and sent messages all over the country, undoubtedly the system would to a large extent supersede the telegraphs and consequently largely diminish the telegraph revenue. Therefore, it is an essential feature of the scheme, if carried out, that the Government should have possession of the trunk wires.

Meanwhile, dissatisfaction with local exchange service mounted. The telephone company was hampered by the reluctance of local authorities to grant wayleaves. Many authorities believed they could offer a better and cheaper service themselves. A deputation from local Chambers of Commerce asked the government to acquire and work the telephone system as a branch of the Post Office. The government agreed that the telephone system was not currently operated to the general benefit of the country, and was unlikely to be so operated while the private monopoly continued. It pleaded that the Post Office was to over-burdened with work to take on the task of running the local exchange system. By the Telegraph Bill of 1899 it authorised the establishment of exchanges by local authorities, to be financed by borrowing against the rates.

Sixty municipalities applied for licences, thirteen licenses were granted and six were in fact taken up. It soon became apparent that the municipalities had underestimated costs and overestimated revenues. The municipal system in Tunbridge Wells lasted for only 14 months before it was sold to the National Telephone Company. Four of the other five systems were sold off in the next 13 years; Hull is the only municipal system remaining to the present day.

It has been claimed that 'there never was a ghost of a chance that municipal enterprise would offer the solution of the problem as a whole' (Lee, 1913, quoted in Kingsbury, 1915, p. 506; see also Meyer, 1907). The real purpose of allowing municipalities to engage in compe-tition with the National Telephone Company was, it is suggested, to reduce the value of that company's assets, and thereby to reduce the purchase price which the government would have to pay when that company's license expired at the end of 1911.

From 1912 onwards, the entire British telephone industry (except Hull) was run by the Post Office. For many years, however, this de-partment system was starved for capital by the Treasury. Back in 1883 the Treasury had cautioned the Post Office 'not to act in anticipation of possible demand', and this restrictive policy continued. It was not until 1961 that growing political pressure, especially from suppliers urging expansion into rural areas, led to the separation of Post Office finances from the Exchequer. It was a further eight years before the Post Office became a public corporation instead of a government department.

The purpose of this brief account is not to criticise the Post Office or Government, although in retrospect many mistakes were clearly made. The purpose is to illustrate the crucial importance of the political pressures which necessarily operate between and on government departments. Is it possible that the policies dictated by the principles of welfare economics can ever be upheld in the face of such pressures?

14.8 The organisation of the British telecommunications equipment supply industry

Telephone administrations throughout the world have made a variety of different arrangements for the supply of equipment. In the USA, the Bell System purchases the bulk of its supplies from its wholly owned subsidiary Western Electric; General Telephone and Electric, too, is its own major supplier. In Britain, the Post Office purchases from a small group of manufacturers, notably General Electric (GEC), Standard Telephone and Cable (STC) (a subsidiary of International Telephone and Telegraph, ITT and Plessey. In the past, there was a formal arrangement for dividing business between specified suppliers. Countries without a local supply industry either grant a monopoly to an overseas supplier (often on condition that he sets up a local plant to manufacture part of the supply), or else invite competitive tenders internationally.

Which of these arrangements is to be preferred? There are obviously advantages and disadvantages associated with each one, and the most appropriate solution will depend on local conditions in each country. There is not space here for an exhaustive analysis, but it will be of interest to examine how the type of ownership of the telephone system appears to have influenced the choice of supply arrangements in the USA and Great Britain, and the policy of the British Post Office towards its suppliers.

In a free-market system, one would expect industries to be organised in such a way as to minimise total costs of production. Specifically, one would expect the equipment supply industry to be integrated with the telephone system if and only if the advantages of co-ordination and security stemming from joint ownership more than offset the advantages for the telephone system of competition between independent suppliers.

For the last half century or so, the Bell system has been the dominant telephone company in the USA. It is privately owned, but regulated with respect to its rates, and to a large extent has been protected from competition. The company has fought vigorously to keep Western Electric as its manufacturing arm, in the face of occasional attempts to enforce a separation (Billingsley, 1973). The fact that Western Electric is not regulated (at least, to the same extent) has led critics to allege that the Bell System makes excessive profits at the manufacturing stage, being able to pass on high prices to the monopoly telephone market. In defence, it has been said that Western Electric's prices are in fact lower than those of independent manufacturers (see the discussion in chapter 13). It is thus not clear whether the existence of regulation is distorting the pattern of organisation of supply. If, as seems likely, the FCC allows a further entry into telecommunications markets, with a resulting increase in competition, it will be interesting to see whether Western Electric begins to lose its favoured position or whether, on the contrary, it is able to expand by selling to competing telephone companies.

In Britain, the Post Office has not been allowed to manufacture its own equipment, though it has powers to do so under the 1970 Post Office Act, and the Post Office itself is reputed to wish to exercise this power. In the past, it experienced difficulties with a purely 'arm's length' relationship with suppliers. Robertson (1947, p. 175) relates how in 1923 the Post Office called together the four manufacturers then working in the automatic field and made an agreement with them each to supply a specified number of lines at a common price. In 1928 a fresh agreement for a period of five years was drawn up, involving an additional manufacturer. The administration of this agreement and its successors necessitated the formation of the Bulk Contracts Committee to decide policy, prices, distribution of contracts and details of technical development. In 1936 a similar arrangement was made to deal with Post Office stores business, involving eight manufacturers. The group of manufacturers who were party to the Agreement came to be known as the 'Telephone Ring'.

In April 1963 the agreements covering cable and loading coils were discontinued; those covering telephone apparatus and exchange equipment were renewed. A modification was inserted into the contracts whereby a higher fixed percentage of equipment could be bought outside the Ring (25% of telephone apparatus and 10% of exchange equipment). In practice, according to Canes (1966, p. 37), very little was purchased outside the Ring or Ring-owned subsidiaries.

When the Post Office was made into a public corporation in 1969, it was expected to act more ruthlessly in seeking new equipment, to shop around to find the best designs, prices and dates of delivery. The old bulk-supply agreements were gradually terminated, although the three major suppliers were alleged by Maddox (1972, p. 222) to have a favoured position, being guaranteed 50% of the orders for exchange equipment according to the formula: Plessey and GEC two-fifths each, STC one-fifth.

At the time of writing there is still, in principle, a policy of competitive purchasing, and international exchanges have been bought from L.M. Ericsson of Sweden. There is also however, an awareness of the desirability of sharing knowledge between manufacturers to produce to a common design, and many people would favour a 'rationalisation' (i.e. merger) of the three British suppliers.

The arrangement of the Ring is frequently defended by the industry. It is argued that, by bulk purchase, and thorough investigation of costs, the Post Office obtained substantially lower prices than had been obtained hitherto (Robertson, 1947, p. 175 and p. 278), or than could be obtained from foreign suppliers. British prices for telephone equipment were alleged to be the lowest in the world; exports were sold for higher prices than the Post Office paid. Cost advantages were obtained by being able to manufacture and maintain standardised equipment to an orderly ordering programme and avoiding duplication of research and development (R&D) activity.

It is further argued that the arrangement actually stimulated and rewarded research and development. The cost of R&D was part of the overheads allowed in the cost investigation. if a manufacturer introduced improved manufacturing methods after costs were investigated, he got the benefit of higher profit during the remainder of the 5-year period and the Post Office obtained the benefit of lower price for the next five years. There was competition in research, both for prestige and because if a manufacturer's proposal was adopted by the Post Office he obtained extra business while he was first in the field and others were not. There was co-operation in development, because all companies had to manufacture (and the Post Office to maintain) from the same drawings.

According to this view, the Ring arrangement was working well, but was broken up for purely political reasons, and this was a tragic mistake. The Post Office is now trying to obtain both competitive purchasing and co-operative development, an arrangement which is not in fact working, and cannot be expected to.

An opposite point of view has also been put forward (Canes, 1966, p.37):

> In 1961 the Committee on Public Accounts criticised the agreement on the ground that absence of competition was harmful. No assurance could be given that the most economic methods were being used: furthermore, the rate of profits earned by the industry were not known to the Post Office. In 1963, the Committee again referred to the ring and emphasised that cost studies of the industry were much too rare.

Pye of Cambridge Ltd., which had to fight bitterly to enter the ring in 1960, also condemned it. Its managing director, Mr. J.R. Brinkley, wrote to *The Times* (24 June 1964):

> .... the present arrangements by which the Post Office purchases its telephone equipment are not in the public interest. These Post Office Bulk Supply Agreements are amongst the most restrictive documents I have ever seen and it is impossible to reconcile them with a modern competitive industrial situation.

The arguments in favour of the bulk-supply agreements are found to be weak:

> Standardisation can be maintained within a competitive framework and moreover is not necessarily wholly of benefit to the consumer. Technological progress may be stifled through insistence on standardisation, since incentive to refine and improve is damped. If some components must remain standard, the GPO could set forth its requirements well before bids were to be tendered. A single buyer is in a strong position to ensure standardisation from its suppliers.
>
> Pooling technical information may have some advantage, but trade publications also assemble and disseminate information. Moreover, competition can be an effective stimulus to research. Easier programming of orders is an absurd objection that reveals a willingness to accept what is offered rather than take the trouble to investigate alternatives.
>
> In 1964 the Post Office Engineering Union published a brochure entitled 'The Telephone Ring' which pointed out that, despite claims to the contrary, British exports of telephone equipment had stagnated. In any event, it is not clear that export potential is primarily a function of the size of the exporting establishment.
>
> If prices have been reduced, it has been despite rather than because of the bulk supply agreements. There is little reason to believe that a cartel will go out of its way to provide minimum prices.

We are not in a position to evaluate the two contrasting arguments here. However, a few remarks may be in order concerning the incentives and pressures operating in and on the Post Office. As a public corporation, it has no private shareholders clamouring for profits, and as a monopoly it has no rivals competing away its business. Although the Government will naturally be concerned to avoid price increases, it seems likely that the Post Office will have less pressure to secure low costs and prices than a private company. One may suppose, in fact, that it would be prepared to sacrifice some reduction in cost to ensure that its suppliers were willing to manufacture to its own specifications equipment of kinds which have extremely limited markets elsewhere.

One would also expect that the government would put pressure on the Post Office to meet other 'national and social needs'. In particular, it seems that the Post Office has been concerned to preserve several independent British manufacturers of telephone equipment, partly to ensure competition, partly for reasons of national prestige and exports, and partly to avoid unemployment at a time when the economy is depressed. Naturally, the manufacturing companies themselves have benefited from and encouraged this policy (though it would seem that they have apparently become increasingly frustrated with the Post Office). That this is by no means a recent change in policy is suggested by some comments by Robertson (1947, p. 176) on the decision to re-equip London with automatic Strowger exchanges on the basis of pooled patents and contract-sharing between manufacturers, which led in turn to the Bulk Contracts Committee:

> It meant that the industry, assured of a steady and controlled domestic market, could afford also to give attention to orders from overseas. It meant that the industry could mature along its own lines, and find its own level. Colonel Purves [British Post Office Chief Engineer] has claimed that 'from the first inception of the Post Office telephone system it has been the policy of the Department to discontinue the purchase of telephone plant from abroad and to encourage the establishment, within the United Kingdom, of adequate manufacturing resources for the supply of all its needs'.

> For the industry, too, played its part. A good deal of the 'encouragement' came, it is worth pointing out, from the industry. Without the London contract the industry in this country would have withered away; but without the industry's insistence and determination not to be sidestepped, there would have been no London contract, so far as British manufacturers were concerned.

An economist cannot say whether the objectives of meeting Post Office design specifications, preserving British manufacturers and avoiding unemployment are worthy ones, nor whether they justify the sacrifices in cost and price which (some allege) are entailed. Objections may be levied against all three objectives. Maddox (1972) sees the history of Post Office decisions on technology as a series of blunders; it may be argued that the British market is no longer big enough to sustain three domestic manufacturers, and that the real competition is on an international level between the UK, Sweden (Ericsson), Germany (Siemens), USA (Western Electric and General Telephone and Electric) and Canada (Northern); finally, it is arguable that artificially avoiding unemployment today only delays inevitable adjustments to a changed world situation, thereby reducing economic growth.

As we have tried to indicate, recent work on property rights and organisational incentives can help us to understand why the Post Office has taken the approach it has. One can also predict that if the Post

Office were not a public corporation, or even if it were subject to more competition by removing its statutory monopoly, it would place less weight on such 'national' and 'social' considerations, and more weight on achieving a low-cost telephone system. Whether such a change in emphasis would be desirable is not for an economist to say, and in any case, we are still somewhat in the dark as to precisely what arrangements between the Post Office and its suppliers would best serve telephone subscribers.

14.9 The rise and fall of marginal-cost pricing

There was never complete agreement among economists about marginal-cost pricing and other aspects of welfare economics, but it did seem that during the 1950s and 1960s the balance of opinion began to swing in their favour. The official endorsement by the British government in the White Paper of 1967 (see chapter 9) was an acknowledgment of the popularity of the doctrine. During the next five years, economists within the civil service and elsewhere, especially within the National Board for Prices and Incomes, devoted considerable energy to working out the implications of this philosophy for a variety of nationalised industries and public enterprises. Yet, with the possible exception of the Electricity Council, there never seemed to be any great enthusiasm on the part of the industries themselves, nor indeed on the part of the government. Over the last decade, the notion of marginal-cost pricing has gradually faded into the background. Why is this?

Although the British government declared in the 1967 White Paper its commitment to the use of marginal-cost pricing and investment appraisal techniques, in practice, it has consistently acted as if various other macroeconomic considerations ought to take priority. The need to stimulate or dampen the economy, to avoid inflation or unemployment, the need for nationalised industries to 'set the lead' for the private sector, the imposition of the price 'freeze' in 1970—these considerations have all prevented any serious application of marginal-cost pricing. Since the sponsoring government departments and the Treasury did not insist upon marginal-cost pricing, the industries were surely justified in feeling that the government did not attach great importance to this doctrine.

In its White Paper on *The Nationalised Industries*, issued in May 1978, the Government acknowledge that the price restraint policies of the early 1970s had mismanaged the Government's responsibility to set a broad financial and economic framework, declared it essential that these mistakes should not be repeated, and reaffirmed the policy of marginal-cost pricing. Within a year, the Government decided to 'freeze' electricity prices, thereby violating once again its declared policy (*Guardian*, 16 March 1979).

For their own part, the managers of a nationalised industry, not being able to withdraw profits for themselves, and not being held to account by profit-conscious shareholders, were likely to seek other sources of satisfaction and to minimise the complaints coming their way. The Post Office might wish to build a technologically advanced telecommunications systems, or it might prefer to avoid as long as possible the complexities of change. It might attempt to build a large empire, to provide secure employment for its own staff and those of its suppliers, or it might seek an easy life by minimising the complaints coming its way. For present purposes the objectives actually pursued are unimportant. The central point being made is that, whatever the objectives of the managers of the nationalised industries, and whatever the objectives of the government, marginal-cost pricing was unlikely to be seen for long as a relevant tool for attaining these objectives. There are precious few votes in efficient resource allocation.

14.10 The ambiguity of marginal-cost pricing

The fact that marginal-cost pricing has not been successful in Britain leads us to question whether it is, in fact, a sound doctrine. The newer views we are currently exploring cast doubt on this. There are basically two objections. The first is that the correct marginal-cost pricing policy cannot be defined objectively; the rule therefore presents inadequate guidance to the nationalised industry and is incapable of being monitored and enforced by a supervisory government agency. The second objection is that marginal-cost pricing, and welfare economics generally, is based on a purely static view of competition, and fails to accommodate the problems of uncertainty, learning and change. We shall explain these points in turn.

If the injunction to set price equal to marginal cost is to mean anything, and to be capable of enforcement, there must exist some well-defined quantity called marginal cost. Now the *concept* of marginal cost is certainly well-defined, but it is not at all apparent *which* marginal cost is the appropriate one to choose. In the conventional timeless, single-product diagrams of economic theory, there is no ambiguity. In practice, as was shown in chapter 5 marginal cost depends on (i) the size of the marginal increment, (ii) the size of the output to which the marginal increment is added, (iii) the date at which the marginal increment is added and (iv) the date at which the calculation is made. To each and every specification of these four 'parameters' corresponds a different marginal cost. Further, since cost refers in principle to foregone future opportunities, it depends not on 'facts' about the past but on conjectures about the future—specifically, about revenues and money outlays associated with the proposed plan and with other, rejected, plans. Here, again, to each and every view about the future corresponds a different collection of marginal costs.

To illustrate, one might argue that the quarterly rental for a telephone in Birmingham in 1979 should be £5, £20 or £50 depending on whether one calculates the marginal cost of (i) a single extra telephone added today to the Birmingham exchange, which presently has spare capacity, or (ii) 50 extra telephones added today to each exchange in the UK, only some of which have spare capacity, or (iii) an increase of 1% in demand over the next 5 years, where that demand is itself expected to grow at 7% per annum and at the same time there are expected to be difficulties with electronic switching which will seriously limit capacity available. An infinite number of other calculations might be made. In each case, rental so determined would be equal to marginal cost. Which price is most reasonable depends on which calculation it is most reasonable to take as a basis, which in turn depends upon how far it is reasonable to differentiate rentals between exchanges of different characteristics, how often it is reasonable to revise the tariff, how fast it is reasonable to expect national income to grow, how difficult it is expected to be to install electronic switching, etc.

It appears, then, that the injunction to set price equal to marginal cost does not specify in an objective way the price to be set. The calculation of cost is necessarily left to the discretion of the industry. This in itself is not necessarily a bad thing. The point is that the injunction to set price equal to marginal cost is an empty one: it amounts to no more than an injunction to set prices in a 'reasonable' way.

From this it follows that the government (or anyone else) cannot check whether price has in fact been set equal to marginal cost, because the latter quantity has not been well-specified. The industry can, if it chooses, interpret *any* price as equal to marginal cost by taking an appropriate view of the future and making an appropriate choice of the four parameters specified above. The government cannot find the industry guilty of not following the marginal-cost pricing rule; all it can do is to disagree with the industry's view of the future or its choice of reasonable parameters. These are matters of opinion, not of fact. If the government enforces its own view, it is not thereby enforcing marginal-cost pricing, but rather taking on itself the responsibility for the industry's pricing policy.

Exactly analogous arguments apply to the rules about investment appraisal: that prospective rate of return should exceed the test discount rate or, equivalently, that net present value at the test discount rate should be positive. Suppose that one manager predicts a fast growth in demand for telephones and believes that technical difficulties associated with electronic switching equipment will soon be solved. Another manager may be less optimistic on both counts. The first manager calculates that demand will be high and marginal costs will be low, generating a high return on capital which will justify a large programme of installing new central office equipment. The second manager believes demand will be lower and costs higher, generating a rate of return which does not justify new investment on such a scale. Both these managers are correctly following the specified principles of investment appraisal, but they make different recommendations because their opinions about the future differ. In such cases, the investment rule says nothing about which alternative ought to be chosen. The government cannot object to either choice on the ground that the rule has not been followed; it can only voice its disagreement with the assumptions used, and if necessary impose its own view.

14.11 The role of entrepreneurial alertness

Turn now to the second objection to marginal-cost pricing, which follows from the Austrian approach to competition. The pricing and investment rules prescribed by the 1967 White Paper were derived from the welfare economic model of static equilibrium. They presume that the relevant products, demands, resources, resource prices and techniques are 'given'. The task of the manager of a nationalised industry is merely to calculate the cheapest way of supplying demand and to set prices accordingly. In practice, however, these data are not given. The task of the manager is precisely to *find out what they are*: to discover how many telephone calls customers want to make, whether they want colour phones, whether electronic switching will prove superior to crossbar, what level of wages will need to be paid to switchboard operators over the next decade, and how this compares with likely installation and operating costs of automatic exchanges etc.

The market mechanism is thus a process of discovery, a process of conjecture and experiment, a search for new and better ways of meeting customers' wishes. If this market mechanism is to be replaced by a nationalised industry (and even if this is done because it is thought that a natural monopoly with only negligible competition would other-wise result), some provision must be made for encouraging this dis-covery process by the industry. The rules about marginal-cost pricing and investment appraisal do not presume to meet this task. They refer to a specified set of alternatives available; they do not prescribe how these alternatives are to be generated. In other words, the concept of entrepreneurial alertness to new opportunities is completely absent from welfare economics.

The 1967 White Paper partly recognised this point:

> To make the best use of resources, it is not enough merely to ensure that prices properly reflect costs, important though this is. Continuous and critical attention has to be paid to costs themselves... Management's task is to seize every opportunity of reducing costs by employing improved methods or techniques. This call for imagination, careful planning and a willingness to experiment. [para. 27]

However, the White Paper provided very little incentive for the indus-tries to indulge in imagination and experiment. If the Post Office is already meeting its prescribed financial target—if, indeed, it is making so much money that refunds have to be made to all subscribers—what incentive is there to be alert to new products, such as a 'Princess' telephone embodying the dial and microphone in a single piece handset? Will not such a device merely be a nuisance to produce and sell, and make customers dissatisfied with existing equipment?

14.12 An alternative framework

This view to which welfare economics leads us is broadly as follows. Telecommunications is largely a natural monopoly which calls for regulation or nationalisation. The industry should then be operated according to the principles of marginal-cost pricing and investment appraisal, as set out in previous chapters of this book. The newer views we have been exploring cast serious doubt on whether these pricing and investment principles will achieve the desired end of efficient resource allocation. More than that, they suggest that telecommunications indus-tries, which in every country are regulated or nationalised, are not in fact operated with the deliberate aim of improving the allocation of resources, nor was this the real reason for regulation or nationalisation in the first place. Indeed, it would not be an exaggeration to say that telecommunication industries were regulated or nationalised precisely to *prevent* the more efficient allocation of resources that would have come about as a result of the competitive process. The government was called on to intervene to protect one or more of the market participants (including, in the case of Britain, the Post Office) from the more unpleasant consequences of the competitive struggle.

The implication of this newer view is that, if it is desired to have a telecommunications industry which is increasingly efficient and alert to customer preferences, then regulation or nationalisation may not be the best framework for the industry. Freedom of entry of new firms is likely to be more effective than the deliberation of a regulatory agency or the exhortations of a sponsoring government department in discovering those products which customers most prefer, and in providing them as soon as possible at the lowest possible price. In other words, the possibility of competition minimises the chance of a good idea being overlooked or suppressed, and at the same time it serves to protect customers from exploitation by incumbent monopolists.

What form competition would take in telecommunications remains to be seen. Recent experience in the USA suggests that many competitors are able and willing to enter into parts of the industry, notably the provision of private lines, high density transmission networks and subscriber equipment. Waverman (1975) has argued that full economies of scale on transmission routes are attained at a level of output which is small compared with the market, so that two or more competing transmission systems could be sustained on major routes. Presumably the market for subscriber equipment would also sustain many companies.

It would seem to be inefficient to have more than one organisation providing local exchange service to each area. But even here it has been suggested that the existence of a single producer *in* the market does not preclude competition *for* the market (Desmetz, 1968). For example, independent television companies in Britain compete at periodic intervals for the right to a temporary monopoly of commercial TV in each area. The difficulties are perhaps somewhat greater with telephone exchanges, since the incoming company would have to take over all the equipment from the outgoing one. Other economists have doubted the feasibility of this approach (Williamson, 1976). Nevertheless, if one were seriously concerned to promote the efficient allocation of resources, this idea would be worth exploring.

Finally, let us suggest a few alternative measures for attaining efficiency within a regulated or nationalised industry. On the one hand, it is necessary to provide incentives to search out new products and lower cost techniques. Insofar as regulated companies can keep part of their profits they have an incentive to do this. Nationalised industries do not have the spur of private ownership; it may be that bonuses linked to financial performance and the right to retain profits to invest as the management desire would be partial substitutes. On the other hand, it is necessary to prevent resources being misallocated by excessive investment or by government intervention motivated by political considerations. This might be done by exposing the industry to competition, by introducing an element of private capital (either debentures or, preferably, equity capital) into the nationalised industries and by insisting that borrowing is charged at market rates.

Government instructions to the industry should be couched in objective rather than subjective terms, i.e. they should be operational. Since injunctions to set price equal to marginal cost or to appraise investments according to a test discount rate are unenforceable, they should be abandoned. In contrast, instructions to break even, to meet a target level of net revenue or to achieve a specified degree of self finance, may easily be monitored. So, too, may instructions to provide specified products at specified prices, if this is thought desirable for social reasons.

Further suggestions are beyond the scope of this book. The main purpose of this chapter has been to alert the reader to some new directions in economic thinking which have important implications for government policy towards telecommunications. The thrust of this work is that the institutional framework within which telecommunications services are provided is of much greater importance than optimal rules of behaviour laid down by economists. A greater interest in the effects of different kinds of institutional framework is therefore likely to characterise future work in telecommunications economics.

REFERENCES

(a) Author list

Aims of Industry: *Posts and telephones: a survey of users' reports on service* (London, no date)

AKNAI, PETER: 'Controlling costs and telephone exchanges', *Telecommunications,* **11,** No. 11, November 1977, pp. 28I–28J

ALCHIAN, A.A.: 'Costs and outputs' *in* ABRAMOVITZ, M. *et al.: The allocation of economic resources,* Stanford University Press, 1959, reprinted as chapter 11 of ALCHIAN, A.A.,: *Economic forces at work* (Liberty Press, Indianapolis, 1977)

ALCHIAN, A.A.: 'Some economics of property rights', *Il Politico,* **30,** No. 4, 1965, pp. 816-29, reprinted in his *Economic forces at work* (Liberty Press, Indianapolis 1977)

ALCHIAN, A.A.: 'Information costs, pricing and resource unemployment', *Economic Inquiry,* **17,** No. 2, June 1969, pp. 109-28

ALCHIAN, A.A., and ALLEN, W.R.: *University economics* (Prentice–Hall, London, 3rd edn. 1974)

ALCHIAN, A.A., and KESSEL, R.A.: 'Competition, Monopoly and the pursuit of money' *in* Universities–National Bureau of Economic Research: *Aspects of labour economics* (Princeton University Press, 1972 pp. 157-75)

ALLEMAN, J.H.: *The pricing of local telephone service.* OT Special Publication 77-14, Office of Telecommunications, US Department of Commerce, April 1977

ANDERBERG, M., and WIKELL, G.: 'Call generation, holding times and traffic load for individual telephone and telex subscriber lines', *Tele,* English edn., XXVIII, no.2, 1976, pp. 37-45

ARTLE, R., and AVEROUS, C.: 'The telephone system as a public good: static and dynamic aspects', *Bell Journal of Economics and Management Science,* **4,** No. 1, Spring 1973, pp. 89-100

AT&T: *The World's telephones.* Published annually by the American Telephone and Telegraph Co., New York

AT&T (American Telephone and Telegraph Company): *Annual Report,* New York

AURAY, R.R.: 'Discussion of "Marginal cost pricing of telephone calls" by S.C. LITTLECHILD', Long Lines Department, American Telephone and Telegraph Company (paper presented at Public Utilities Seminar, Darmouth College, September 1969)

AURAY, R.R.: 'Elasticity of demand for communications services' *in Economics of regulated communications industry in the age of innovations,* (Boston, 1970)

AURAY, R.R.: 'Consumer response to changes in interstate MTS rates' *in* TREBING, H.M. (Ed.): *Assessing new pricing concepts in public utilities,* MSU Public Utilities Papers, 1978, pp. 47-82

AVERCH, H., and JOHNSON, L.L.: 'Behaviour of the firm under regulatory constraint', *American Economic Review,* **52,** no. 5, December 1962, pp. 1053-69

BAER, W.S., and MITCHELL, B.M.: 'The economic impact of competition on an independent telephone company', *Public Utilities Fortnightly,* 23 October 1975, pp. 1-7

BAILEY, ELIZABETH E.: *Economic theory of regulatory constraint* (Lexington Books, DC Heath & Co., 1973)

BAILEY, ELIZABETH E., and WHITE, L.J.: 'Reversals in peak and off-peak prices' *Bell Journal of Economics and Management Science,* **5,** no.1, Spring 1974, pp. 75-92

BALDWIN, G.G.C.: *The history of the telephone in the United Kingdom* (Chapman & Hall, London, 1925)

BAUMOL, W.J.: *Economic theory and operations analysis* (Prentice –Hall, Englewood Cliffs, NJ, 1961, 3rd edn., 1972)

BAUMOL, W.J.: Testimony of William J Baumol'. FCC Docket 16258, Bell Exhibit 26, 31 May 1966

BAUMOL, W.J.: 'Reasonable rules for rate regulation: plausible policies for an imperfect world' *in* PHILLIPS, A., and WILLIAMSON, OLIVER E.., (Eds.): *Prices: issues in theory, practice, and public policy* (University of Pennsylvania Press, Philadelphia, 1967)

BAUMOL, W.J.: 'Testimony of Dr. William J. Baumol'. FCC Docket 18128/ 18684, Bell Exhibit 12, July 1970

BAUMOL, W.J.: 'Rate making: incremental costing and equity considerations' *in Essays on public utility pricing and regulation,* TREBING, HARRY M., (Ed.): MSU Public Utilities Studies, Institute of Public Utilities, Michigan State University, 1971

BAUMOL, W.J.: 'Quasi-optimality: the welfare price of a nondiscriminatory price system' *in* WENDERS, JOHN T. (Ed.): *Pricing in regulated industries: theory and application,* Mountain States Telephone & Telegraph Co., Denver, 1977*a*

BAUMOL, W.J.: 'On the proper cost tests for natural monopoly in a multiproduct industry', *American Economic Review,* **67,** no.5, December 1977*b,* pp. 809-822

BAUMOL, W.J., BAILEY, E.E., and WILLIG, R.D.: 'Weak invisible hand theorems on pricing and entry in a multi-product natural monopoly', *American Economic Review,* **67,** June 1977, pp. 350 – 365

BAUMOL, W.J., and BRADFORD, D.F.: 'Optimal departures from marginal cost pricing', *American Economic Review,* **60,** June 1970, pp. 265 – 83

BAUMOL, W.J., ECKSTEIN, OTTO and KAHN, A.E.: 'Competition and monopoly in telecommunications services' Testimony submitted in FCC docket 18920 (Specialised common carriers), 2 December 1970

BAUMOL, W.J., and KLEVORICK, A.K.: 'Input choices and rate-of-return regulation: an overview of the discussion', *Bell Journal of Economics and Management Science,* **1** no. 2, Autumn 1970, pp. 162 – 190

BEBEE, E.L., and GILLING, E.J.W.: 'Telecommunications and economic development: a model for planning and policy making', *ITU Telecommunications Journal,* **43,** no. 8, August 1976, pp. 537-43

Bell Telephone Laboratories Staff: *Transmission systems for communications* Bell Telephone Laboratories, 4th ed, 1970)

BECH, N.K.: 'Analysis of demand for telephone stations', CCITT, no.18-E, August 1967

BECH, N.K.: 'Price influence on the demand for telephone calls', CCITT, August 1967

BENNETT, A.R., : *The telephone systems of the continent of Europe* (Longmans, 1895)

BENSON, D.L., and SHURROCK, C.R.J.: 'Econometric modelling to investigate telephone exchange economics', *Br. Post Office Electrical Engineers Journal,* **67**, 1974

BETTERIDGE, W.W.: Supplement to testimony, FCC Docket 19129, Phase II, November 1973

BIALA, J.: *Transfers between public telephone exchanges.* Ministry of Communications, Tel Aviv, Israel, June 1975

BILLINGSLEY, J.R.: 'Values of vertical integration in the Bell System'. Paper presented at Sixth Annual Seminar on Economics of Public Utilities, Wisonsin Bell Co., March 1973 (published as a pamphlet by Western Electric)

BISHOP, P.,: 'Queueing theory in the Ada telephone system'. UK West Midlands Gas Board, Technical Report OR 21, February 1971

BLACK, S.K., and TRYON, P.V.: 'Increased telephone installation rates: a statistical analysis of Colorado's 1972 rate change'. US Department of Commerce, Office of Telecommunications, June 1976, OT 76-90, US Government Printing Office

BOHM, E.: 'Analytical models for forecasting the demand for telephone connections'. Bonn: Federal Ministry of Posts & Telecommunications, Bonn, W. Germany. Paper presented at Sixth International Teletraffic Conference, Munich, 9–15 September 1970

BØ, K., GANSTAD, O., and KOSBERG, J.E.: 'Some traffic characteristics of subscriber categories and the influence from tariff changes'. Paper presented at the Eighth International Teletraffic Congress, Melbourne, Australia, November 1976

BOITEUX, M.: 'La tarification au coût marginal et les demandes aléatoires', *Cahiers du Seminaire d'Econometrie,* **1**, 1951

BOITEUX, M.: 'La tarification de demandes en point: application de la théorie de la vente au coût marginal', *Reve Générale de l'Electricité,* **58**, August 1949, 321–40. Translated as 'Peak-load pricing', *Journal of Business,* **33**, April 1960, pp. 157–79. Reprinted *in* NELSON, J.R. (Ed.): *Marginal cost pricing in practice* (Prentice–Hall, Englewood Cliffs, 1974)

BOITEUX, M.: 'Sur la gestion des monopoles publics astreints a l'equilibre budgetaire', *Econometrica,* 24 January, 1956, pp. 22-40. Translated as 'On the management of public monopolies subject to budgetary constraints', *Journal of Economic Theory,* **3**, September 1971, pp. 219–40

BOLTER, W.G.: 'The FCC's selection of a "proper" costing standard after fifteen years--what can we learn from docket 18128?' *in* TREBING, H.M. (Ed.): *Assessing new pricing concepts in public utiilities,* MSU Public Utilities Papers, 1978, pp. 333–373

BONBRIGHT, J.C.: *Principles of public utilities rates* (Columbia University Press, New York, 1961)

BONILLA, A.: *Rural telecommunications in Costa Rica.* ICE, San Jose, Costa Rica, October 1977 (paper presented at Intelcom 77, Atlanta, Ga.)

BORTS, G.H.: 'Factors affecting the growth of the communication industry'. Paper presented at a seminar on the Economics of the Regulated Communications Industry in the Age of Innovation, sponsored by New England Telephone & Telegraph Co. and held at Dartmouth College, New Hampshire, 15–19 June 1964

BOWER, L.L.: 'Telecommunications market demand and investment requirements', *Telecommunication Journal,* **39**, no. 3, March 1972, pp. 177-181

BOWERS, D.A., and LOVEJOY, W.F.: 'Disequilibrium and increasing costs: a study of local telephone service', *Land Economics,* **41**, no. 1, February 1965, pp. 31-40

BRAEUTIGAM, R.R.: 'An investigation of cross subsidy, industry structure and pricing, with applications to data under voice'. Staff Research Paper OTP-SP-14, Office of Telecommunications Policy, Executive Office of the President, Washington, DC20504, July 1973*a*

BREAUTIGAM, R.R.: Addendum to 'An examination of the burden test'. August 1973*b*

BRAEUTIGAM, R.R.: 'More on the surplus pricing method for multiproduct transmission systems: DUV costs and shifting demands'. Studies in Industry Economics 47, (Discussion Paper), Stanford University, July 1974

BRAEUTIGAM, R., OWEN, B., and ULEN, T.: 'Rate structure and open entry: the problem of regulating competition'. Memorandum 166, Stanford University, Center for Research in Economic Growth, Stanford, California, February 1974

BRANDON, BELINDA B., and BRANDON, PAUL: *The effects of the demographics of individual households on their telephone usage in Chicago* (to be published)

British Post Office: 'Report of a study of marginal costs of exchange lines and calls services', Telecommunications Finance Department, London, October 1971

British Post Office: 'Report and Accounts', (HMSO, London, Annual)

BROOKS, J.: *Telephone: the first hundred years* (Harper & Row, New York, 1976)

BROWN, R.: *Telecommunications: the booming technology* (Aldus Books, London, 1969)

BUCHANAN, J.M.: *The demand and supply of public goods* (Rand McNally, Chicago, 1968)

BUCHANAN, J.M.: *Cost and choice* (Markham Publishing Co., Chicago, 1969)

BUCHANAN, J.M.: 'Consumerism and public utility regulation' *in*PHILLIPS, C.F. Jr. (Ed.): *Telecommunications, regulation and public choice* (Washington and Lee University, Va. 1975), pp. 1-22

BUCHANAN, J.M., and THIRLBY, G.F.: *LSE essays on cost* (Weidenfeld & Nicolson, London, 1973)

BUCHANAN, J.M., and TOLLISON R.D. (Eds.): *Theory of public choice* (University of Michigan Press, Ann Arbor, 1972)

BUCHANAN, J.M., and TULLOCK, G.: *The calculus of consent* (University of Michigan Press, Ann Arbor, 1962)

CANES, M.: *Telephones—public or private?:* Hobart Paper 36, Institute for Economic Affairs, London, June 1966

CANTWELL, J., and HONEYMAN, N.: 'Manpower planning for telephone development'. Government Operations Research Unit, Department of the Public Service, Dublin, Ireland (paper presented to the International Conference on Telecommunications Economics, University of Aston in Birmingham, May 1974)

CAPELLO, F., and SANNERIS, A.: 'Studio economico delle accessibilita in una central telefonica automatica, *Alta Frequenza,***25**,nos.3-4,1975/6, pp. 304-18

CARLIN, A., and PARK, R.E.: 'Marginal cost pricing of airport runway capacity', *American Economic Review*, **60**, no. 3, June 1970, pp. 310-319

CARLTON, LYNNE:'Sensitizing telephone rates', *Telephony*, 15 September 1975

CARTER, C.F. (Chairman): *Report of the Post Office Review Committee.* Cmnd. 6859, HMSO, London, July 1977 (also Appendix, Cmnd. 6954, November 1977)

CAÑAS, A.F.: *System development patterns & policies: the case of Cost Rica – its plans and achievements.* ICE, San Jose, Costa Rica, October 1977 (paper presented at Intelcom 77, Atlanta, Ga.)

CCITT: *Economic studies at the national level in the field of telecommunications,* (1964-68). ITU, Geneva, July 1968

CHADDHAR, R.L., and SHARAD S. CHITGOPEKAR: A generalisation of the logistic curves and long-range forecasts (1966-1991) of residence telephones', *Bell Journal of Economics and Management Science,* **2**, Autumn 1971, pp. 542-560

CHAMBERS, D., and CHARNES, A.: 'Intertemporal analysis and optimization of bank portfolios', *Management Science,* 7, no. 4, July 1961

CHAPUIS, R.J.: 'Technology and structures: man and machines', *Telecommunications Policy,* March 1978

CHARNES, A., COOPER, W.W., and MILLER, M.H.: 'Application of linear programming to financial budgeting and the costing of funds', *Journal of Business,* **32**, No. 1, January 1959, pp. 20-46

CHERRY, C.: *World communication – threat or promise? A socio-technical approach,* John Wiley, (revised edn.), Chichester, 1978

CHERRY, E.C.: 'The conceivable future of telecommunications', *Trans. S. Afr. Inst. Electr. Engrs.,* **49,** P.8, August 1958, pp. 271-286

CLARKE, A.C.: *Voice across the sea* (William Luscombe, London, 1st dn., 1958, revised, 1974)

CLEMENS, E.W.: 'Price discrimination in decreasing cost industries', *American Economic Review,* **31,** 1941, pp. 794-802

CLEMENS, E.W.: *Economics and public utilities* (Appleton-Century-Crofts Inc., New York, 1950)

CLOS, C., and WILKINSON, R.I.: 'Dialling habits of telephone customers', *Bell Syst. Tech. J.,* **31,** 1952, pp. 32-67

Cmnd. 3437: *Nationalised industries: a review of economic and social objectives,* HMSO, London, 1967

Cmnd. 7131: *The nationalised industries,* HMSO, London, March 1978

Cmnd. 7292: *The Post Office,* HMSO, London, July 1978

COASE, R.H.: 'The marginal cost controversy', *Economica,* 13 August 1946, pp. 169-82

COASE, R.H.: 'The Federal Communications Commission', *Journal of Law and Economics,* October 1959, pp. 1-40

COASE, R.H.: 'The Interdepartment Radio Advisory Committee', *Journal of Law and Economics,* October 1962, pp. 17-47

COASE, R.H.: 'Evaluation of public policy relating to radio and television broadcasting: social and economic issues', *Land Economics,* **41,** 1965, p. 161

COASE, R.H.: 'The economics of broadcasting and government policy', *American Economic Review Papers and Proceedings,* 56, 1966, p. 440

COASE, R.H.: 'The theory of public utility pricing and its application', *Bell Journal of Economics and Management Science,* **1,** no. 1, Spring 1970, pp. 113-128

COHEN, G.: 'Experimenting with the effect of tariff changes on traffic patterns'. Eighth International Teletraffic Congress, November 1976 (to be published)

COHEN, G.: 'Measured rates versus flat rates: a pricing experiment'. Paper presented at Fifth Annual Conference on Telecommunications Policy Research, Airlie House, Virginia, March 1977

COLLIER, M.E.: 'Economic planning of transmission systems', *Electrical Communication,* **48**, nos. 1, 2, 1973, pp. 189-92

COWAN, D.D. and WAVERMAN, L.: 'The interdependence of communications and data processing: issues in economics of integration and public policy', *Bell Journal of Economics and Management Science,* **2**, no. 2, Autumn 1971, pp. 657-677

COX, K.A.: 'An appraisal of regulatory pricing policies in communications' *in* TREBING, H.M. (Ed.): *Essays on public utility pricing and regulation,* MSU Public Utilities Studies, Institute of Public Utilities, Michigan State University, 1971

CRACKNELL, D.R.: 'New econometic models for full rate trunk calls'. Report 89, Statistics and Business Research Department, British Post Office, April 1977

CRAVEN, J.: 'On the choice of optimal time periods for a surplus-maximizing utility subject to fluctuating demand', *Bell Journal of Economics and Management Science,* **2**, no. 2, 1971, pp. 495-502

CRINER, J.C.: 'Telecommunications resale: a policy analysis', *Telecommunications Policy,* September 1977*a*

CRINER, J.C.: 'What is valued added network service?', *Telecommunications,* October 1977*b*

CRINER, J.C.: 'Resale and shared use: winds of change', *Telecommunications,* **11**, November 1977*c*

CRINER, J.C.: 'VANS—current regulatory issues', *Telecommunications,* **12**, no. 7, July 1978

CRUTCHLEY, E.G.: *G.P.O.* (Cambridge University Press, 1938)

CUNNINGHAM, P.A., and LURIN, E.S.: 'The future of value added network services', *Telecommunications,* **12**, no. 7, July 1978

DANIELS, R.W.: 'Efficiency and performance in the British public enterprise sector 1960-74'. Paper presented to Fifth Conference on Industrial Structure, European Association for Research in Industrial Economics, Nurnberg, September 1978

DASGUPTA, A.K., and PEARCE, D.W.: *Cost benefit analysis: theory and practice* (Macmillan, 1972)

DAVIS, B.E., CACCAPPOLO, G.J., and CHAUDRY, M.A.: 'An econometric planning model for American Telephone and Telegraph Company', *Bell Journal of Economics and Management Science,* **4**, no. 1, Spring 1973, pp. 29-56

DAVIES, D.W., and BARBER, D.L.A.: *Communication networks for computers* (Wiley, 1973)

DE ALESSI, L.: 'An economic analysis of government ownership and regulation: theory and the evidence from the electric power industry', *Public Choice,* **19**, Autumn 1974

DE BUTTS, J.D.: 'An unusual obligation'. Speech before the Annual Convention of the National Association of Regulatory Utility Commissioners, Seattle, Washington, 20 September 1973

DE BUTTS, J.D.: 'The US communications consumer and FCC policies: the AT&T position on the proposed legislation', *Telecommunications Policy,* March 1977

DE MONTBRIAL, T., and MULLER, J.C.: *Tarification du Telephone*, June 1969

DE VANY, ECKERT, MEYERS, O'HARA and SCOTT: 'A property system for market allocation of the elctromagnetic spectrum: a legal-economic-engineering study '*Standford Law Review*, **21**, 1969, p. 1499

DEMSETZ, H.: 'Why regulate utilities?', *J. Law & Econs.*, **11**, April 1968, pp. 55-66

DESCHAMPS, P.: 'A short term pricing model for telephone calls with applications to the Belgian Network' Mimeograph, September 1972

DESCHAMPS, P.: 'The demand for telephone calls in Belgium 1961 – 1969'. Paper presented at the International Conference in Telecommunications Economics, University of Aston in Birmingham, May 1974

DESCHAMPS, P.J.: 'Guidelines for incorporating some aspects of current tele-traffic theory in a pricing model with endogenous point-to-point blocking probabilities'. Mimeograph, October 1974

DESCHAMPS, P.J.: 'Second-best resource allocation in telephone networks with alternate-routing facilities, endogenous shortage probabilities and pricing constraints: existence and computation'. Carnegie-Mellon University, School of Urban and Public Affairs, February 1975

DESCHAMPS, P.J.: *Second-best pricing with variable product quality: a quantitative decision rule and an existence proof.* Université Catholique De Louvain, Faculté Des Sciences Economiques, Sociales et Politiques, Nouvelle Serie 128. 1976

DESCHAMPS, P.J.: 'Pricing for congestion in telephone networks: a numerical example' *in* KARAMARDIAN, S. (Ed.): *Fixed points: algorithms and applications* (Academic Press, 1977), pp. 473-94

DICKENSON, C.R.: 'Telecommunications in developing countries—the relation to the economy and the society'. Proceedings of Intelcom 77, Atlanta, Ga., October 1977

Dittberner Associates Inc: *Internconnection: an economic impact analysis.* Consultancy report prepared for Office of Telecommunications Policy, Executive Office of the President, Washington, DC

DOBELL, A.R., TAYLOR, L.D., WAVERMAN, L., LIV, T.H., and COPELAND, M.D.G., 'Telephone communications in Canada: demand, production and investment decisions' *Bell Journal of Economics and Management Science*, **3**, no. 1, Spring 1972, pp. 175-219

DOHERTY, A.N.: 'Econometric estimation of local telephone price elasticities', *in* TREBING, H.M. (Ed.): *Assessing new pricing concepts in public utilities*, MSU Public Utilities Papers, 1978, pp. 101-129

DREW, J.M.: 'Econometric models of international communications'. Statistics & Business Research Department, T2 Division, British Post Office Central Headquarters Report 28, October 1973

DREW, J.M., and CRACKNELL, D.R.: 'An econometric model of cheap rate trunk calls'. Report 73, Statistics and Business Research Department, British Post Office, November 1976

DREZE, J.A.: 'Some post war contributions of French economicsts to theory and public policy with special emphasis on problems of resource allocation', *American Economic Review*, **54**, no. 4, Pt. 2, 1964, pp. 1-64

DUNN, D.M., WILLIAMS, W.H., and SPIVEY, W.A.: 'Analysis of prediction of telephone demand in local geographical areas', *Bell Journal of Economics and Management Science*, **2**, no. 2, Autumn 1971, pp. 561-576

EADS, G.C.: 'Telecommunications regulation: what can we learn from other regulated industries?' *in* PHILLIPS, C.F., Jr. (Ed.): *Telecommunications, regulation and public choice* (Washington & Lee University, Virgina, 1975), pp. 57-77

EASTMOND, L.E.: 'Testimony of L.E. Eastmond', FCC Docket 18128/18684, Bell Exhibit 3, 1 April 1970

ECKERT, R.D.: 'Spectrum allocation and regulatory incentives' *in Conference on communications policy research: papers and proceedings,* Office of Telecommunications Policy, 1972

The Economist: ' Will Britain join the citizens' band?', 15 October 1977, pp. 82-3

ELLINGHAUS, W.M.: 'Testimony of William M.. Ellinghaus', FCC Docket no. 16250, Bell Exhibit 46, 15 September 1967

ELLIS, L.W.: 'The law of the economies of scale applied to telecommunications system design', *Electrical Communication,* **50,** no. 1, 1975, pp. 4-19

ELLIS, L.W.: 'The impact of economies of scale on the investment decision' *in* POLISHUK and O'BRYANT (Eds.): *Exposition proceedings, telecommunications & economic development,* papers presented at the first Internatioal Telecommunication Exposition, Atlanta, Ga. 9-15 October, 1977

ELROM, H.: 'Telecommunications and the national economy' Mimeograph, Ministry of Communications, Tel Aviv, Israel, 1972

ENDE, A.H.: 'International telecommunications: dynamics of regulation of a rapidly expanding service', *Law and Contemporary Problems,* **34,** no. 2, pp. 389-417, 1969

EVERS, R.: 'Analysis of traffic flows on subscriber-lines dependent on time and subscriber-class'. Eighth International Teletraffic Congress, Paper 345, Melbourne, Australia, November 1976

FARRELL, M.J.: 'In defence of public utility price theory', *Oxford Economic Papers,* N.S. **10,** 1958, pp. 109-123, reprinted with amendments in TURVEY (Ed.): *Public enterprise—selected readings* (Penguin Books, Harmondsworth, 1968)

FAULHABER, G.: 'Cross subsidization: pricing in public enterprise', *American Economic Review,* **65,** December 1975, pp. 966-77

FELDSTEIN, M.S.: 'Equity and efficiency in public sector pricing: the optimal two-part tariff', *Quarterly Journal of Economics,* **86,** no. 2, May 1972, pp. 175-187

FERGUSON, C.E.: *The neoclassical theory of production and distribution* (Cambridge University Press, 1971)

FERGUSON, C.E., and GOULD, J.P.: *Microeconomic theory,* R.D. Irwin, Illinois, 1st edn. 1966, 4th edn. 1975)

FLOOD, J.E. (Ed.): *Telecommunication networks* (Peter Peregrinus, London, 1975)

Florida Public Service Commission, Docket 74805-TP (CR), Order 7018, 1975

FLOWERS, T.H.: *Introduction to exchange systems* (Wiley, 1976)

FOCHLER, W.C.: 'Life-line: welfare pricing of local telephone service' *in* TREBING, H.M. (Ed.): *Assessing new pricing concepts in public utilities,* MSU Public Utilities Papers, 1978, pp. 463-474

FOX, R.W.B.: 'A review of the short-term forecasting model for total trunk calls'. Statistics & Business Research Department, British Post Office Central Headquarters, Report 27, April 1973

FROGGATT, A.M.: *Testimony* and *Supplemental testimony,* FCC Docket 16258, Bell Exhibits 24 and 24A, 31 May and 29 July 1966

FROGGATT, A.M.: 'Incremental costing in practice' *in* TREBING, H.M. (Ed.): *Essays on Public Utility Pricing and Regulation,* MSU Public Utilities Studies, Michigan State University, East Lansing, 1971, pp. 151-66

FURUBOTN, ERIK G., and PEJOVICH, SVETOZAR: 'Property rights and economic theory: A survey of recent literature', *Journal of Economic Literature,* **10,** December 1972, pp. 1137-62

FUSS, M., WAVERMAN, L. and DENNY, M.: 'Productivity, technological change and returns to scale in Canadian telecommunications'. Paper presented at Seventh Annual Telecommunications Policy Research Conference, Skytop, Pa., April 1979

GABEL, R.: *Development of separations principles in the US telephone industry,* Institute of Public Utilities, Michigan State University, 1976

GABEL, R.: 'The early competitive era in telephone communication, 1893-1920', *Law and Contemporary Problems,* **34,** no. 2, Summer 1969, pp. 340-360

GABEL, R.: 'Barriers to entry and boundaries between regulated and unregulated telecommunication services: the equipment market'. Paper presented at the Telecommunications Policy Research Conference, Airlie House, Virginia, 17-19 April 1974

GALE, W.A.: Material from Bell Telephone Laboratories Memorandum embodied in Petition by Pacific North West Bell Telephone Company to State of Washington, concerning Intrastate Message Toll Service, Cause U-75-40, Exhibit 66, witness L.K. Baumgartner, 12 January 1976

GAMBLE, R.B.: 'VAN services in the US', *Telecommunications,* **12,** no. 7, July 1978

GARFINKEL, L.: 'Usage sensitive pricing: studies of a new trend', *Telephony,* 10 February 1975

GELLERMAN, R.: 'Measuring the benefits of telecommunications' *in* POLISHUK and O'BRYANT (Eds.)' *Exposition Proceedings, Telecommunications & Economic Development,* First International Telecommunications Exposition, Atlanta, Georgia, 9-15 October 1977

General Post Office (London), Statistics & Business Research Department: 'Econometric models of the telecommunications industry'. Report 1: 'The growth of business connexions', May 1968

GILL, SIR FRANK: 'Engineering economics', *I.E.J.* 90 part 1 no. 33, September 1943.

GORDON, M.J.: *The investment financing and valuation of the corporation,* (Irwin, Illinois, 1962)

GRAAFF, J. de V.: *Theoretical welfare economics* (Cambridge University Press, 1957)

GRANDMONT, Jean-Michel: 'Temporary general equilibrium theory', *Econometrica,* **45,** no. 3, April 1977, pp. 535-72

GRANT, E.L.: *Principles of engineering economics* (Ronald Press, New York, 3rd edn. 1950)

GRAVELLE, H.S.E.: 'A note on telephone rentals: comment', *Applied Economics,* **4,** no. 2, 1972, pp. 235-238

GREGORY, P.: *Telephones for the elderly,* Occasional Papers on Social Administration 53 (Bell, London 1973)

HAHN, F.H.: *On the notion of equilibrium in economics* (Cambridge University Press, 1973)

HAI, S.: 'Market structure and the development of a modern telecommunications network'. Draft paper, Ministry of Communications, Tel Aviv, Israel, 1976

HALL, J.K.: 'Differential telephone rates', *Journal of Land and Public Utility Economics,* 1932

Jack Faucett Associates: 'Improved rate structure in telecommunications'. OTP-SE-72-114, US Department of Commerce, Office of Telecommunications Policy, US Government Printing Office, Washington, DC, 1972

HAMSHER, D.H. (Ed.): *Communication system engineering handbook* McGraw–Hill, 1967)

HARKNESS, R.C.: *Telecommunications substitutes for travel.* US Department of Commerce, Office of Telecommunications, Washington, DC, December 1973

HASSELWANDER, A.C.: 'The cost of competition in the terminal equipment market' *in* TREBING, H.M. (Ed.): *Assessing new pricing concepts in public utilities,* MSU Public Utilities Papers, 1978, pp. 321-333

HAYEK, F.A.: 'Economics and knowledge' *Economica,* **4** 1937, pp. 33-54, reprinted in his *Individualism and economic order*

HAYEK, F.A.: *Individualism and economic order* (University of Chicago Press, 1948)

HAYEK, F.A.: *The counter-revolution of science* (The Free Press of Glencoe, New York, 1952)

HAZLEWOOD, A.: 'Optimum pricing as applied to telephone service', *Review of Economic Studies,* **18,** 1950-51, pp. 67-78, reprinted with amendments as 'Telephone service' *in* TURVEY, R., (Ed.): *Public enterprise, Selected Readings* (Penguin Books, Harmondsworth, 1968), pp. 237-257

HILLS, M.T., and EVANS, B.G.: *Transmission systems* (Allen & Unwin, 1973)

HIRSHLEIFER, J.: 'Peak loads and efficient pricing: comment', *Quarterly Journal of Economics,* **72,** August 1958, pp. 451-62

HIRSHLEIFER, J., DEHAVEN, J.C., and MILLIMAN, J.W.: *Water supply* (University of Chicago Press, 1960)

HODES, D.A.: 'Impact of business cycles on telephone gain', *Telephone Engineer and Management,* October 1963

HOLCOMBE, A.N.: *Public ownership of telephones on the continent of Europe* (Houghton Mifflin Co., Boston & New York, 1911)

HOPLEY, J.K.: 'The response of local telephone usage to preak pricing *in* TREBING, H.M., (Ed.): *Assessing new pricing concepts in public utilities,* MSU Public Utilities Papers, 1978, pp. 3-12

HORN, C.E.: 'Factors affecting telephone pricing' *in* WENDERS, JOHN T. (Ed.): *Pricing in regulated industries: theory and application,* Mountain States Telephone & Telegraph Co., Denver, 1977

HOUTHAKKER, H.S.: 'Electricity tariffs in theory and practice', *Economic Journal,* **61,** March 1951, pp. 1-25

HOTELLING, H.: 'The general welfare in relation to problems of taxation and of railway and utility rates', *Econometrica,* **6,** July 1938, pp. 242-69

IEE Conference Publication: 'Private electronic switching systems', no. 163, 1978

Instituto Costarricense de Electricidad (ICE): *Telecommunications feasibility study stage IV,* San Jose, Costa Rica, 1973

IRWIN, M.R.: *The telecommunications indsutry: integration versus competition* (Praeger, New York, 1971)

IRVIN, G.: *Modern cost-benefit methods* (Macmillan, London, 1978)

Israel, Ministry of Communications: 'Measuring productivity in a communications system' (Working paper), Jerusalem 1974

JENSEN, A.: *Moe's principle: an econometric investigation intended as an aid in dimensioning and managing telephone plant*, The Copenhagen Telephone Company, Copenhagen 1950

JENSON, A.: 'The applicability of decision theory in the planning and operation of telephone plant' *Teleteknick*, **1**, 1957, pp. 126-129

JIPP, A.: 'Wealth of nations and telephone density, *ITU Telecommunications Journal*, July 1963

JOEL, A.E.: *Electronic switching: central office systems of the world* (IEEE Press, 1976)

JOHNSON, L.L.: 'Communications satellites and telephone rates: problems of government regulation'. Prepared for National Aeronautics and Space Administration, Memorandum RM-2845-NASA, October 1961, The Rand Corporation, Santa Monica, California

JOHNSON, L.L.: 'Joint cost and price discrimination: the case of communications satellites' *in* SHEPHERD, W.G., and GIES, THOMAS G., (Eds.) *Utility regulation: new directions in theory and policy* (Random House, New York, 1966)

JOHNSON, L.L.: 'Technological advance and market structure in domestic telecommunications', *The American Economic Review*, Papers and Proceedings of the 32nd meeting of the American Economic Association, New York, 28-30 December 1969

JOHNSON, L.L.: 'Behaviour of the firm under regulatory constraint: a reassessment', *American Economic Review*, **63**, May 1973, pp. 90-7

JOHNSON, L.L.: 'The problem of regulated specialized communications common carriers', Rand Corporation, Santa Monica, California, 1976

JOHNSON, L.L.: 'A review of the FCC and AT&T positions', *Telecommunications Policy*, March 1977

JOHNSON, N.: 'Towers of Babel: the chaos in radio spectrum utilization and allocation', *Law and Contemporary Problems*, **34**, no. 3, Summer 1969

JOHNSTON, J.: *Econometric methods* (McGraw‾Hill, 2nd edn. 1963

JOLLEY, E.H.: *Introduction to telephony and telegraphy* (Pitman, 1968)

JORDAN, W.A.: 'Producer protection, prior market structure and the effects of government regulation', *Journal of Law and Economics*, **15**, April 1972, pp. 151-76

KAHN, A.E.: 'The graduated fair return: comment', *American Economic Review*, **58**, March 1968, pp. 170-73

KAHN, A.E.: *The economics of regulation*, Vols. I and II (Wiley, New York 1970-71)

KAHN, A.E., and ZIELINKSI, C.A.: 'New rates structure in communications', *Public Utilities Fortnightly*, 25 March 1976*a*

KAHN, A.E., and ZIELINKSKI, C.A.: Proper objectives in telephone rate structuring', *Public Utilities Fortnightly*, 8 April 1976*b*

KAO, K.C., and COLLIER, M.E.: 'Fibre-optic systems in future telecommunication networks', *Telecommunications*, April 1977, pp. 25-32

KAY, J.A.: 'Pricing and investment criteria in public utilities: the case of telephone service'. Unpublished MA dissertation, University of Edinburgh, 1968

KEMP, G.S.: 'VANS in Europe', *Telecommunications*, **12**, no. 27, July 1978

KENWARD, M.: 'Tomorrow's telephones: the technical debate', *New Scientist*, **53**, no. 785, 2 March 1972, pp. 468-74

KEYES, N., and GERLA, M.: 'Hybrid packet and circuit switching', *Telecommunications*, July 1978

KINGSBURY, J.E.: *The telephone and telephone exchanges* (Longmans, Green, London, 1915)

KIRZNER, I.M.: *Competition and entrepreneurship* (University of Chicago Press, 1973)

KIRZNER, I.M.: *The perils of regulation: a market-process approach*. Occasional Paper, Law and Economics Center, University of Miami, School of Law, 1978

KLASS, M.W., and SHEPHERD, W.G. (Eds.): *Regulation and entry*, MSU Public Utilities Papers, 1967

KLEIN, B.H., GOLDSEN, J.M., LIPSON, L.S., MECKLING, W.H., MOORE, F.T., and REIGER, S.H.: 'Communications satellites and public policy: an introductory report'. Memorandum RM-2925-NASA, The Rand Corporation, Santa Monica, California, December 1971

KLEVORICK, A.K.: 'The graduated fair return: a regulatory proposal', *American Economic Review*, **56**, June 1966, pp. 577-84

KLEVORICK, A.K.: 'The behaviour of a firm subject to stochastic regulatory review', *Bell Journal of Economics and Management Science*, **4**, 1973, pp. 57-88

KNIGHT, F.H.: 'Some fallacies in the interpretation of social cost', *Quarterly Journal of Economics*, **38**, 1924, pp. 528-606

KNIGHT, N.V.: 'Depreciation and service life of telecommunication plant', *Br. PO Elect. Engrs. J.*, Paper 209, February 1955a

KNIGHT, N.V.: 'Economic principles of telecommunications plant provision' *Br. PO Elect. Engrs. J.*, Paper 210, November 1955b

KOHLMEIER, L.M. Jr.: *The regulators* (Harper & Row, New York, 1969)

KOLM, S.C.: L'économie normative des services de masse',*Rev, Econ. Politique,* (Paris), **80,** Jan/Feb. 1970, pp.61-84

KOLM, S.C.: 'Service optimal d'une demande variable et le prix de l'incertitutde' *Rev. Economique*, (Paris), **21**. no.2, March 1970, pp. 243-71

KOPINSKI, J.W.: 'The USE international record carriers' *Telecommunications*, May 1977

KORTANEK, K.O.: 'A linear programming supply model for switched probabilistic communications networks'. Mimeograph, December 1977

KOSBERG, J.E., GAUSTAD, O., and KRISTIAN, B.: 'Some traffic characteristics of subscriber categories and the influence from tariff changes', Eighth International Teletraffic Congress, Melbourne, Australia, November 1976

KOTOWITZ, Y., and WAVERMAN, L.: 'A model of the demand for telephone services in Canada'. Mimeograph, June 1973

KRAEPELIEN, H.Y.: 'The influence of telephone rates on local traffic'. Paper read at the Second International Teletraffic Congress, The Hague, 7-11, July 1958

KROES, J.L. de: 'Calculation of the number of direct function lines in telephony', *Commun. News*, **12**, June 1952, pp. 132-43

KROES, J.L. de: 'Optimum grouping in uniselector trunking', *Commun. News*, **14**, January 1954, pp. 70-6

LAGO, A.M.: 'Demand forecasting models of international telecommunications and their policy implications', *Journal of Industrial Economics*, **19**, no. 1, November 1970, pp. 6-21

LARKIN, E.P.: 'The brave new world of telecommunications' *in* PHILLIPS, C.F. (Ed.): *Monopoly in the domestic communications industry*: (Washington & Lee University, 1974), p.60

LAYARD, R. (Ed.): *Cost benefit analysis* (Penguin Modern Economic Readings, Harmondsworth, 1972)

LEE,: *Economics of telegraphs and telephones* (Pitman, London, 1913)

LELAND, H.E., and MEYER, R.A.: 'Monopoly pricing structures with imperfect discrimination', *Bell Journal of Economics, 7*, no. 2, Autumn 1976, pp.449-62

LESSING, L.: 'Cinderella in the sky', *Fortune, 76*, October 1967, pp. 131-208

LEVIN, H.J.: 'Federal control of entry in the broadcast industry', *Journal of Law and Economics*, October 1962, pp. 49-67

LEVIN, H.J.: 'New technology and the old regulation in radio spectrum management', *American Economic Review*, May 1966, pp. 339-49

LEVIN, H.J.: 'The radio spectrum resource', *Journal of Law and Economics*, October 1968, pp. 433-501

LEVIN, H.J.: 'Spectrum allocation with markets', *American Economic Review*, Papers and Proceedings, May 1970

LEVY-LAMBERT, H.: 'Tarification des services a qualité variable–application aux péages de circulation', *Econometrica, 36*, no. 3-4 July-October 1968, pp. 564-574

LEWIN, L. (Ed.): *'Telecommunications: an interdisciplinary survey* (Artech House, Dedham, MA, 1978)

LEWIS, BEN W.: 'Emphasis and misemphasis in regulator policy' *in* SHEPHERD, W.G., and GIES, THOMAS G. (Eds.): *Utility regulation: new directions in theory and policy* (Random House, New York, 1966)

LEWIS, C.D.: 'Demand analysis and short-term forecasting', chapter 10 *in* LITTLECHILD, S.C. (Ed.): *Operational Research for managers* (Philip Allan, Oxford, 1977)

LEWIS, W. ARTHUR: 'The two-part tariff', *Economica, 8*, 1946, pp. 169-82, reprinted in revised form as chapter 2 of his *Overhead costs*

LEWIS, W. ARTHUR: *Overhead costs* (Unwin, London, 1949)

LINDER, R.W.: 'The development of manpower and facilities planning for airline telephone reservations offices', *Operational Research Quarterly, 20*, no. 1, March 1969

LIPSEY, RICHARD G.: *An introduction to positive economics* (Weidenfeld & Nicolson, 1963, 3rd edn. 1971)

LIPSEY, R.G., and LANCASTER, K.J.: 'The general theory of the second best' *Review of Economic Studies, 24*, no. 1, 1956

LITTLE, I.M.D., and MIRRLEES, J.A.: *Project appraisal and planning for develping countries* (Heinemann, 1974)

LITTLE, I.M.D., and SCOTT, M.F.G. (Eds.): *Using shadow prices* (Heinemann, 1976)

LITT'ECHILD, S.C.: 'A note on telephone rentals', *Applied Economics, 2*, no. 2, May 1970*a*, pp. 73-4

LITTLECHILD, S.C.: 'A game-theoretic approach to public utility pricing', *Western Economic Journal, 8*, no. 2 June 1970*b*, pp. 162-66

LITTLECHILD, S.C.: 'Marginal-cost pricing with joint costs', *Economic Journal, 80*, June 1970*c*, pp. 323-335

LITTLECHILD, S.C.: 'Peak-load pricing of telephone calls', *Bell Journal of Economics and Management Science, 1*, no. 2, Autumn 1970*d*, pp. 191-210

LITTLECHILD, S.C.: 'A state-preference approach to public utility pricing and investment under risk', *Bell Journal of Economics and Management Science,* **3,** no.1, Spring 1971, pp. 340-345

LITTLECHILD, S.C.: *The telephone system in Costa Rica.* Report to IBRD Public Utilities Division, April 1973*a*

LITTLECHILD, S.C.: 'Myopic investment rules and toll charges', *Journal of Transport Economics and Policy,* **7,** no. 2, May 1973*b*

LITTLECHILD, S.C.: 'Optimal arrival rate in a simple queueing system', *Int. J. Prod. Res.,* **12,** no. 3, 1974, pp. 391-397

LITTLECHILD, S.C.: 'Common costs, fixed charges, clubs and games', *Review of Economic Studies,* **42,** no. 1, 1975*a*, pp. 117-124

LITTLECHILD, S.C.: 'Two-part tariffs and consumption externalities', *Bell Journal of Economics,* **6,** no. 2, Autumn 1975*b*, pp. 661-70

LITTLECHILD, S.C.: 'Organisation and performance: an international comparison of telecommunications systems'. Memorandum of evidence submitted to the Carter Committee on the British Post Office, 16 November 1976 (also presented to the First International Telecommunication Exposition, Atlanta, Georgia, October 1977)

LITTLECHILD, S.C.: 'The role of consumption externalities in the pricing of telephone service' *in* WENDERS, J.T. (Ed): *Pricing in regulated industries: theory and application* (Mountain States Telephone & Telegraph Co., Denver, 1977)

LITTLECHILD, S.C.: *The fallacy of the mixed economy,* Hobart Paper 80, Institute of Economic Affairs, London, June 1978

LITTLECHILD, S.C.: 'Distributional aspects of telecommunications policies'. Paper presented at Seventh Annual Telecommunications Policy Research Conference, Skytop, PA., April 1979

LITTLECHILD, S.C., and ROUSSEAU, J.J.: 'Pricing of a U.S. telephone company, *Journal of Public Economics,* **4,** 1975, pp. 35-36

LONGLEY, H.A.: 'Some grade-of-service problems'. 6th International Teletraffic Congress, 1970, pp. 522/1-522/4

LONNQVIST, I.: 'Econometric models for investment profitability'. Financial Department, Swedish Telecommunications Administration, 1975

LONNSTROM, S., MARKLUND, F., and MOO, I.: *A telephone development project* (Telefonaktiebolaget, Stockholm, L.M. ERICSSON, 2nd edn. April 1965, 3rd edn. [with P.BIRD], June 1967)

LONNSTROM, S., MOO, I., and ERICSSON, L.M.: 'Economic aspects in telephone systems' *in* POLISHUK and O'BRYANT (Eds.): *Exposition Proceedings: Vol. 1, Telecommunications & Economic Development.* First international Telecommunications Exposition, Atlanta, Georgia, 9-15 October 1977, pp. 87-92

LORENZ, C.: 'International comparisons of common carrier rates and efficiency: the quest for objectivity', *Telecommunications Policy,* March 1978

LORIE, J.H., and SAVAGE, L.J.: 'Three problems in rationing capital', *Journal of Business,* **28,** no. 4, October 1955

LOWRY, E.D.: 'Justification for regulation: the case for natural monopoly', *Public Utilities Fortnightly,* **28,** 8 November 1973, pp. 1-7

LOWRY, E.D.: 'The Demand for Telecommunications Services: A Survey', paper presented at 1976 Telecommunications Policy Research Conference, Airlie House, Virginia, April 21-4, 1976

MCELROY, F.W.: 'Economies of scale in the provision of service, a network analysis'. Mimeograph, Georgetown University, 1975. Abstract of paper presented to Telecommunications Policy Research Conference, Airlie House, Virginia, April 1975, printed in *Report on proceedings*, OWEN. M. BRUCE (Ed).

MACGREGOR, MARY E.: 'What contribution can the theory of optimal pricing make to the construction of a better tariff structure for a telephone system, with particular reference to the present position of long distance calls in the telecommunications business of the British Post Office'. Unpublished essay, Clare College, Cambridge, 1978

MCKIE, JAMES W.: 'Regulation and the free market: the problem of boundaries', *Bell Journal of Economics and Management Science,* **1,** no.1, Spring 1970, pp. 6-26

MADDOX, BRENDA: *'Beyond Babel: new directions in communications'* (Andre Deutsch, London, 1972)

MANDANIS, GEORGE P.: 'An empirical analysis of economies of scale and specialization in communications' *in* TREBING, H.M. (Ed.) *New dimensions in public utility pricing,* Michigan State University Public Utilities Studies, East Lansing, 1976, pp. 324-88

MANTELL, LEROY H.: 'An econometric study of returns to scale in the Bell System'. Abstract of paper presented to Telecommunications Policy Research Conference, Airlie, Virginia, April 1975*a*, printed in *Report on proceedings,* OWEN, BRUCE M. (Ed.)

MANTELL, LEROY H.: 'Returns to scale in telecommunications'. Office of Telecommunication, Department of Commerce and Office Telecommunications Policy, Executive Office of the President mimeograph, 1975*b*

MANUS, PETER C.: 'An interrelationship between demography and telecommunications' *in* PHILLIPS, C.F. (Ed.): *Telecommunications, regulation and public choice* (Washington & Lee University, Virginia, 1975), pp. 79-99

MARCHAND, M.G.: 'Welfare pricing of a randomly rationed comodity under a budget constraint: the case of telephone services'. CORE Discussion Paper 6711, August 1967, Center for Operations Research and Econometrics, Louvain, Belgium

MARCHAND, M.G.: 'Priority pricing with application to time-shared computers'. Working Paper 247, Center for Research in Management Science, University of California, Berkeley, March 1968

MARCHAND, M.G.: 'The economic principles of telephone rates under a budgetary constraint', *Review of Economic Studies,* **40,** no. 4, October 1973

MARTIN, J.: *Future developments in telecommunications* (Prentice–Hall, 2nd edn., 1977)

MATHEWSON, G. FRANKLIN, and QUIRIN, G. DAVID: 'Metering costs and marginal cost pricing in public utilities, *Bell Journal of Economics and Management Science,* Spring 1972

MAYSTON, D.J.: 'Optimal licensing in public sector tariff structures' *in* PARKIN, M., and NOBAY, A.R. (Eds.) *Contemporary issues in economics* (Manchester University Press, 1975)

MEADE, J.E.: 'Price and output policy of state enterprices', *Economic Journal* **54,** 1944, pp. 321-328

MEADE, J.E.: *The intelligent radical's guide to economic policy: the mixed economy* (George Allen & Unwin, London 1975)

MECKLING, W.H.: 'Management of the frequency spectrum'. *Wash. ULQ,* 1968

MEEK, R.L.: 'An application of marginal cost pricing—the Green Tariff. Part I: Theory, Part II: Practice', *Journal of Industrial Economics,* November 1963

MELODY, W.H.: 'Testimony of William H. Melody'. FCC Docket 16258, FCC Exhibit 52, 25 November 1968

MELODY, W.H.: 'Interservice subsidy: regulatory standards and applied economics' *in* TREBING, H.M. (Ed.): *Essays on public utility pricing and regulation,* MSU Public Utilities Studies, Institute of Public Utilities, Michigan State University, East Lansing, 1971, pp. 167-210

MENGER, C.: *Principles of economics* (1871. Translated by DINGWALL, J. and HOSELITZ, B. Illinois Free Press, Glencos, 1950)

MERRETT, A.J., and SYKES, A,: *The finance and analysis of capital projects* (Longmans, London, 1963)

MERRIMAN, J.H.H.: 'Engineering innovation in a service industry: Post Office telecommunications' (President's Inaugural Address), *Proceedings of the Institution of Electrical Engineers,* **122,** no. 1, January 1975

MEYER, CHARLES W.: 'The cost function for local telephone service: increasing or decreasing'. Dissertation submitted to the Faculty of Philosophy of the Johns Hopkins University, Baltimore, Maryland, 1961

MEYER, CHARLES W.: 'Urban growth and the cost of local telephone service' *Land Economics,* **42,** no. 3, 1962

MEYER, CHARLES W.: 'Marginal cost pricing of local telephone service', *Land Economics,* No. , 1966, pp. 378-383

MEYER, H.R.: *Public ownership and the telephone in Great Britain* (Macmillan, London, 1907)

MILLER, ROGER LEROY: *The economics of European social issues,* (Harper & Row, London, 1975)

MILLS, G.: 'Public utility pricing for joint demand involving a durable good', *Bell Journal of Economics and Management Science,* **7,** no. 1, Spring 1976, pp. 299-307

MILLWARD, R.: *Public expenditure economics: an introductory application of welfare economics* (McGraw—Hill, London, 1971)

MINA, R.R.: 'Traffic standards for grade of service' *Telephony,* June 1974, pp. 33-37

VON MISES, L: *Human action* (Henry Regnery Co., Chicago, 1st edn. 1949, 3rd revised edn. 1963)

MITCHELL, BRIDGER M.: 'Pricing policies in selected European telephone systems'. Discussion paper dp/78-47, International Institute of Management, Berlin, June 1978. Also issued as Rand Corporation Paper P6169, July 1978*a* (paper presented at Sixth Annual Telecommunication Policy Research Conference, Airlie House, Virginia, 10-13 May 1978)

MITCHELL, BRIDGER M.: 'Optimal pricing of local telephone service', *American Economic Review,* **68,** no. 4, September 1978*b*, pp. 517-537

MITCHELL, B.M.: 'Telephone call pricing in Europe: localizing the pulse'. Rand Research Memorandum P-6215, January 1979*a*, forthcoming in WENDERS, J. (Ed): *Pricing in regulated industries, Vol. II*

MITCHELL, B.M.: 'Economic aspects of measured-service telephone pricing', Rand Memorandum P-6321, March 1979*b*

MITCHELL, DAVID C.: 'The impact of competition in communications on settlements and separations' *in* TREBING, H.M. (Ed.): *Assessing new pricing concepts in public utilities,* MSU Public Utilities Papers, 1978, pp. 373-387

The Financial Times: 'Mobile communications' 14 February 1979

MORGAN, CHARLES STILLMAN: *Regulation and management of public utilities* (Houghton & Mifflin Co., Boston, 1923)

MORGAN, T.J.: *Telecommunications economics* (MacDonald, 1st edn. 1958; Technicopy Ltd., Stonehouse, Glos, 2nd edn. 1976)

MOSSLER, ELISABETH: 'Peak-load pricing for trunk calls', *Tele*, English Edn., no. 1, 1977, pp. 49-55

MULLER, JURGEN: Economics of regulation: some issues for Europe'. Preliminary draft for the 5th EARIE Conference, Nurnberg, W. Germany, 13-15 September 1979

MUNASINGHE, MOHAN: 'The economic costs of electric power outages and the optimum level of reliability'. Unpublished report by the World Bank, 1977

MUNBY, DENYS: *Transport* (Penguin Modern Economics, Harmondsworth, 1968)

MURRAY, SIR EVELYN: *The Post Office* (London, Putnam, 1927)

MYSKJA, A., and WALMANN, O.O.: 'An investigation of telephone user habits by means of computer techniques'. Paper presented to the 6th International Symposium on Human Factors in Telecommunications, Stockholm, 1972

MYSKJA, A., and WALMANN, O.O.: 'A statistical study of telephone traffic data, with emphasis on subscriber behaviour'. Seventh International Teletraffic Congress, Paper 132, Stockholm, June 1973

NAOE, SHIGEHIKO: 'An analysis of the interrelations between international telecommunications and economy'. Research Institute of Telecommunications and Economics, Tokyo, 1972. Paper presented at International Conference on Telecommunications Economics, University of Aston in Birmingham, May 1974

NAOR, P.: 'The regulation of queue size by levying tolls', *Econometrica*, 37, no. 1, January 1969, pp. 15-24

NATA (North American Telephone Association): 'NATA position statement', *Telecommunications*, 10, no.10, October 1976

National Board for Prices and Incomes: 'Post Office charges'. Report 58, HMSO, Cmnd, 3574, 1968

NELSON, BOYD L.: 'Econometrics and applied economic analysis in regulatory decisions', *Law and Contemporary Problems: Communications: Part I*, 34, no. 2, Spring 1969, pp. 330-339

NELSON, Boyd L.: 'Problems in the analysis of telecommunications demand' *in* TREBING, H.M. (Ed.): *New dimensions in public utility pricing,* (Michigan State University, 1976) pp. 307-320

NELSON, JAMES R. (Ed.): *Marginal cost pricing in practice* (Prentice–Hall, Englewood Cliffs, NJ, 1964)

NELSON, JAMES R.: 'Pricing and resource allocation: the public utility sector' *in* SHEPHERD, W.G., and GIES, THOMAS G. (Eds.): *Utility regulation: new directions in theory and policy* (Random House, New York, 1966)

NEWSTEAD, I.A.: 'The choice of grade of service standards'. Paper presented to the 3rd International Teletraffic Congress, Paris, 1961

NEWSTEAD, I.A.: 'Telecommunications' future share of national resources' *in* POLISHUK and O'BRYANT (Eds.): *Exposition Proceedings, Telecommuni-*

cations and Economic Development. Papers presented at the first International Telecommunications Exposition, Atlanta, Georgia, 9-15 October 1977

NEWSTEAD, I.A., VASUDEVEN, C.P., DICKENSON, C.R., and JENNINGS, J.H.: *Telecommunications handbook, Part I: An outline of telecommunications, Part II: Telecommunications in developing countries* (International Bank for Reconstruction and Development, Washington DC, 1974)

NG, Y., and WEISSER, M.: 'Optimal pricing with a budget constraint—the case of the two part tariff', *Review of Economic Studies,* **51,** no.127, July 1974, pp. 337-345

NICHOLS, R.T.: 'Submarine telephone cables and international telecommunications'. Memorandum RM-3472-RC, The Rand Corporation, Santa Monica, California, February 1963

NOLL, ROGER G.: *Reforming regulation* (Brookings Institution, Washington DC, 1971)

OI, W.Y.: 'A Disneyland dilemma: two-part tariffs for a Micky Mouse monopoly' *Quarterly Journal of Economics,* **85,** no. 1, February 1971

PALM, C.: 'Methods of judging the annoyance caused by congestion', *Tele,* no.2, 1953, pp. 1-20

PANZAR, J.C., and WILLIG, R.D.: 'Free entry and the sustainability of natural monopoly', *Bell Journal of Economics and Management Science,* **8,** no.1, Spring 1977, pp. 1-22

PEACOCK, ALAN T., and ROWLEY, CHARLES K.: 'Welfare economics and the public regulation of natural monopoly' *Journal of Public Economics,* **1,** August 1972, pp. 227-244

PEARCE, ALAN: 'Washington players in the telecommunications game', *Telecommunications,* **12,** no. 1, January 1978*a*

PEARCE, ALAN: 'Washington waves' *Telecommunications,* **12,** no.4, April 1978*b*

PEARCE, ALAN: 'NTIA—Washington's latest bureaucracy', *Telecommunications,* **12,** no. 6, June 1978*c*

PEARCE, ALAN: 'Regulating communications' *Telecommunications,* **12,** no. 9, September 1978*d*

PEARCE, ALAN: 'Divest manufacturing?' *Telecommunications,* **12,** no.10, October 1978*e*

PEARCE, D.W.: *Cost benefit analysis* (Macmillan, 1971)

PECK, MERTON J.: 'The single-entity proposal for international telecommunications', *American Economic Review,* **60,** May 1970, pp. 199-201

PELTZMAN, S.: 'Pricing and public and private utilities: electric utilities in the U.S.', *Journal of Law and Economics,* **14,** no. 1, April 1971

PERL, LEWIS J.: *Economic and demographic determinants of telephone availability.* National Economic Research Associates, Inc., 5 April 1975, filed by AT&T Company, FCC Docket 20003, Bell Exhibit 21

PHILLIPS, C.F., Jr.: *The economics of regulation* (Irwin, Homewood, Ill, 1969)

PHILLIPS, C.F. (Ed.): *Competition and monopoly in the domestic communication industry,* (Washington & Lee University, Lexington, Va., 1974)

PHILLIPS, C.F., Jr. (Ed.): *Telecommunications, regulation, and public choice.* Papers presented at a symposium sponsored by Washington & Lee University in co-operation with the Chesapeake and Potomac Telephone Companies (Washington & Lee University, 1975)

PIGOU, A.C.: *The economics of welfare* (MacMillan, London, 1920, 4th edn. 1932)

POOLE, JAMES: 'System X—whose hand on the plug?' and 'Can we make it, can we sell it?' *The Sunday Times,* 30 April and 7 May 1979

POSNER, MICHAEL V.: *Post Office orders for telecommunications switching equipment.* Report to the Secretary of State for Industry, London May 1977

POSNER, R.A.: 'Natural monopoly and its regulation', *Stanford Law Review,* **21,** pp. 548-643, February 1969

POSNER, RICHARD A.: 'Taxation by regulation', *Bell Journal of Economics and Management Science,* **2,** Spring 1971, pp. 22-50

POUSETTE, TOMAS: 'The demand for telephones and telephone service in Sweden'. Presented at the European Meetings of the Econometric Society, Helsinki, August, 1976

President's Task Force on Communications Policy: *Final report.* 14 August 1967, US Government Printing Office, Washington DC 20402, 1968

PRESSMAN, I.: 'A mathematical formulation of the peak-load pricing problem', *Bell Journal of Economics and Management Science,* **1,** no.2, Autumn 1970, pp. 304-326

PYATT, G.: 'Some economics of a public utility'. Centre of Industrial Economic & Business Research, University of Warwick, Coventry CV4 7AL, Paper 28, September 1972

PYE, ROGER: 'Focus on Europe: the British Post Office—an extensive monopoly', *Telecommunications Policy,* September 1977

RAPP, Y.: 'The economic optimum in urban telephone network problems', *Ericsson Technic,* no. 49, 1950, pp. 1-132

RAPP, Y.: 'Extension of telephone plant with regard to the value of subscribers' time', *Ericsson Technics,* no. 1, 1961, pp. 5-29

REEVES, H.S.V.: 'System concepts, applications and economics of digital systems', *Telecommunications,* April 1977, pp. 66 A-E

REIGER, S.H., NICHOLS, R.T., EARLY, L.B., and DEWS, E.: 'Communications satellites: technology, economics and system choices', Memorandum RM-3487-RC. The Rand Corporation, Santa Monica, California, February 1963

RENTON, R.W.: *The international telex service* (Pitman, 1974)

REUSS, ROBERT P.: 'Nationalization: is the climate right for U.S. utilities?' *Telephone Engineer and Management,* 1 October 1975, pp. 99-102

ROBBINS, L.: *An essay on the nature and significance of economic science* (Macmillan, London, 1932, 2nd edn. 1935)

ROBERTS, J.H., and WARD, G.B.: 'A comparative study of short term forecasting techniques used in planning a telephone network'. Paper presented to the International Conference on Telecommunications Economics, University of Aston, Birmingham, May 1974

ROBERTSON, J.H.: *The story of the telephone* (Pitman, London, 1947)

RODENBURG, N.: 'Alternative routing of junction traffic', *Comm. News,* **10,** no. 30, 1949

ROHLFS, J.: 'A theory of interdependent demand for a communications service', *Bell Journal of Economics and Management Science,* **5,** no. 1, Spring 1974, pp. 16-37

ROHLFS, J.: 'Economically-efficient Bell-System pricing'. Economics Discussion Paper 138, Bell Laboratories, January 1979

ROOS, I., NORRBY, D., and LEIJON, I.: 'Telephone rates in various countries', *Tele,* **27,** special issue, 1976

ROSE, JOSEPH R.: 'Telephone rates and cost behaviour', *Land Economics,* August 1950, p. 250

Roskill Commission on the Third London Airport: *Report,* HMSO 1971

RUBIN, M., and HALLER, C.E.: *Communication switching systems* (Reinhold, 1966)

RUGGLES, N.: 'Recent developments in the theory of marginal cost pricing' *Review of Economic Studies,* **17,** 1949-50, pp. 107-126. Reprinted *in* TURVEY (Ed.): *Public enterprise: selected readings* (Penguin Books, Harmondsworth, 1968)

SAMUELS, J.M., and WILKES, F.M.: *Management of company finance* (Nelson, London, 1971)

SANDBERG, IRWIN G.: 'Two theorems on a justification of the multi-service regulated company', *Bell Journal of Economics and Management Science,* **6,** Spring 1975, pp. 346-356

SAUNDERS, R.J., and WARFORD, JEREMY J.: 'Telecommunications pricing and investment in developing countries. PU Report PUN 30, International Bank for the construction & Development, Washington DC, June 1977. (paper presented at International Telecommunications Exposition, Atlanta, October 1977, and published in Proceedings thereof)

SCHERER, F.M.: *Industrial market structure and economic performance* (Rand McNally & Co., Chicago, 1970)

SCHKOLNICK, RAUL: *The economics of telephone networks: the case of Costa Rica.* Inter-American Development Bank, Washington DC, January 1976

Select Committee on Nationalised Industries: *The Post Office,* Vol. 1 — Report and Proceedings, House of Commons Paper 340, HMSO, London, 24 February 1967

Select Committee on Nationalised Industries: Session 1967-68, *Ministerial control of the nationalised industries,* Vol. I — Report and Proceedings, House of Commons Paper 371-I, HMSO, London, 24 July 1968

SHACKLE, G.L.S.: *Decision, order and time in human affairs* (Cambridge University Press, 1961, 2nd edn. 1969)

SHACKLE, G.L.S.: *Epistemics and economics* (Cambridge University Press, 1972)

SHARKEY, W.W., and TELSER, L.G.:'Suportable cost functions for the multi-product firm'. Unpublished paper, Bell Laboratories, 1977

SHEAHAN, JOHN B.: 'Competition versus regulation as a policy aim for the telephone equipment industry'. Unpublished Ph.D. dissertation, Harvard University, 1951, pp. 90-91

SHEAHAN, JOHN: 'Integration for exclusion in the telephone equipment industry' *Quarterly Journal of Economics,* May 1956, p. 251

SHEPHERD, W.G.: 'Utility growth and profits under regulation' *in* SHEPHERD, W.G., and GIES, THOMAS G. (Eds.): *Utility regulation: new directions in theory and policy* (Random House, New York, 1966a)

SHEPHERD, W.G.: 'Residence expansion in the British telephone system', *Journal of Industrial Economics,* **14,** July 1966b

SHEPHERD, W.G.: 'The competitive margin in communications' *in* CAPRON, W.M. (Ed.): *Technological change in regulated industries* (Brookings Institution, Washington DC, 1971), pp. 86-122

SHEPHERD, W.G.: 'Entry and communications' *in* PHILLIPS, C.F. Jr. (Ed.): *Competition and monopoly in the domestic telecommunications industry* (Washington & Lee University, Lexington, Va. 1974), pp. 37-54

SHORT, J.A.: 'Long range social forecasts: congestion and quality of service'. Long Range Intelligence Bulletin 10, Post Office Telecommunications, Ref. LRIB 0010/SOF, April 1976

SIMPSON, FLOYD R.: 'Cost trends in the telephone industry', *Journal of Land and Public Utilities Economics,* **21,** August 1945, pp. 286-294

SIMPSON, W.F.: 'Business demand for telephone connexions'. Statistics & Business Research Department, Model Building Branch, British Post Office, October 1969

SIMPSON, W.F.: 'Import restrictions and return on capital' *in* POLISHUK & O'BRYANT (Ed.): *Exposition Proceedings,* Vol. 1, Telecommunications & Economic Development. Papers presented at the first International Tele-communication Exposition, Atlanta, Georgia, 9-15 October 1977, pp. 99-103

SMEED, R.J. (Chairman): *Road pricing: The economic and technical possibilities* (HMSO, London, 1964)

SMITH, ADAM: *The wealth of nations* (London, 1776)

SMITH, S.F.: *Telephony and telegraphy* (Oxford, 2nd edn., 1974)

SNOW, MARCELLUS S.: 'Investment cost minimization for communications satellite capacity' *Bell Journal of Economics and Management Science,* **6,** no.2, 1975, pp. 642

SPARROW, L.B.: 'Manning the telephone enquiry bureau at West Midlands Gas' *in* LITTLECHILD, S.C. (Ed.): *Operational research for managers,* (Phillip Allan Publishers Ltd., Oxford, 1977)

SQUIRE, L.: 'Some aspects of optimal pricing for telecommunications' *Bell Journal of Economics and Management Science,* **4,** no.2, Autumn 1973, pp. 515-525

SQUIRE, L., and VAN DER TAK, H.G.: *Economic analysis of projects* (Johns Hopkins University Press, 1975)

STANNARD, R.: 'The impact of imposing directory assistance charges' *in* TREBING, H.M. (Ed.): *Assessing new pricing concepts in public utilities,* MSU Public Utilities Papers, 1978, pp. 82-101

STEINER, PETER O.: 'Peak loads and efficient pricing', *Quarterly Journal of Economics,* **71,** November 1957, pp. 585-610

STEINER, PETER O.: 'Peak load pricing revisited' *in* TREBING, H.M. (Ed.): *Essays on public utility pricing and regulations* (Michigan State University, 1971)

STERN, CARL: 'Price elasticity of local telephone service demand', *Public Utilities Fortnightly,* 4 February 1965, pp. 24-34

STIGLER, G.J.: 'The economics of information', *Journal of Political Economy,* **69,** no. 3, June 1961

STIGLER, G.J.: 'The theory of economic regulation', *Bell Journal of Economics and Management Science,* **2,** no. 1, Spring 1971, pp. 3-21

STIGLER, G.J., and FRIEDLAND, C.: 'What can regulators regulate? The case of electricity', *Journal of Law and Economicts,* **5,** October 1962, pp.1-16

STONE, K. AUBREY: *I'm sorry, the monopoly you have reached is not in service,* (Ballantine Books Inc., New York, 1973)

SUDIT, E.F.: 'Additive nonhomogeneous production functions in telecommunications', *Bell Journal of Economics and Management Science,* **4,** no. 2, Autumn 1973, pp. 499-514

SYSKI, R.: *Introduction to congestion theory in telephone systems* (Oliver & Boyd, Edinburgh and London, 1960)

TAYLOR, LESTER: 'The demand for communications service in 1980' *Staff Paper 1*, TB 184412, Appendix B, National Technical Information Service, Springfield, Va.

THAYER, G.N.: 'Testimony of Gordon N. Thayer'. FCC Docket 16258, Bell Exhibit 23, 31 May 1966

THIRLBY, G.F.: 'Economists' cost rules and equilibrium theory', *Economica*, May 1960

THOMPSON, HOWARD E. and TIAO, GEORGE C.: 'Analysis of telephone data: a case study of forecasting seasonal time series', *Bell Journal of Economics and Management Science*, **2**, no. 2, Autumn 1971, pp. 515-541

The Times: Communications, 4 April 1979

TOMASEK, O.: 'Forecasting techniques used in Bell Canada' *in* POLISHUK & O'BRYANT (Eds.): *Exposition Proceedings*, Vol. 1, Telecommunications & Economic Development. Papers presented at the first International Telecommunication Exposition, Atlanta, Georgia, 9-15 October 1977, pp. 727-733

TREBING, H.M.: 'Toward an incentive system of regulation', *Public Utilities Fortnightly*, **72**, 18 July 1963, p. 22

TREBING, H.M.: (Ed.): *Performance under regulation* (Michigan State University, East Lansing, 1968)

TREBING, H.M.: 'Common carrier regulation–the silent crisis. Communications: Part 1', *Law and Contemporary Problems*, **34**, no. 2, 1969, pp. 299-330

TREBING, H.M. (Ed.): *Essays on public utility pricing and regulation*, MSU Public Utilities Studies, Institute of Public Utilities, Michigan State University, 1971

TREBING, H.M.: Realism and relevance in public utility regulation' *Journal of Economic Issues*, **8**, no. 2, June 1974

TREBING, H.M. (Ed.): *Assessing new pricing concepts in public utilities*, Michigan State University Public Utilities Papers, 1978

TREBING, H.M., and HOWARD, R. HAYDEN (Eds.): *Rate of return under regulation: new directions and perspectives*, MSU Public Utilities Studies, Institute of Public Utilities, Division of Research, Graduate School of Business Administration, Michigan State University, Michigan, 1969

TREBING, H.M. and MELODY WILLIAM H.: 'Entry conditions in telecommunications' *in* KLASS, M.W., and SHEPHER, W.G. (Eds.): *Regulation and Entry*, MSU Public Utilities Paper 1967, chapter 5

TROXEL, C.E.: 'Rivalry and pricing', *Economics of Public Utilities and Land Economics*, 1947

TROXEL, C.E.: 'Telephone regulation in Michigan' *in* SHEPHERD, W.G., and GIES, THOMAS G. (Eds.): *Utility regulation: new directions in theory and policy*, (Random House, New York, 1966)

TURNER, W.M.: 'Forecasting the growth of business telephone connexions'. Statistics & Business Research Department T2 Division, British Post Office Central Headquarters, Report 26, March 1973

TURNER, W.M.: *CEPT study of the growth of the telephone service.* Report by United Kingdom Telecommunications Headquarters Marketing Department, British Post Office, February, 1975

TURVEY, R.: 'Peak-load pricing' *JPE.*, **76**, no.1, Jan-Feb 1968*a*, pp. 101-113

TURVEY, R. (Ed.): *Public enterprise: selected readings* (Penguin Books, Harmondsworth, 1968*b*)

TURVEY, R.: *Optimal pricing and investment in electricity supply* (Allen & Unwin, London, 1968*c*)

TURVEY, R.: 'Marginal cost', *Economic Journal,* **79**, no. 314, June 1969, pp. 282-299

TURVEY, R.: 'Public utility pricing and output under risk: comment' *American Economic Review,* **60**, no. 3, June 1970, pp. 485-486

TURVEY, R.: *Economic analysis and public enterprises.* (Allen & Unwin, London, 1971)

TURVEY, R.: 'Externality problems in telephone pricing'. Paper presented at the International Conference in Telecommunications Economics, University of Aston in Birmingham, May 1974

TURVEY, R., and ANDERSON, D.: *Electricity economics—essays and case studies* (Published for the World Bank, Johns Hopkins University Press, 1977)

TYLER, R.M.: 'The value of time as a factor in telecommunications planning: some issues for discussion'. Paper presented to a seminar at the Civil Service College, June 1973

TYLER, R. MICHAEL: 'The evaluation of telecommunications programmes and projects in less-developed countries' *in* POLISHUK & O'BRYANT (Eds.): *Exposition Proceedings,* Vol. 1, Telecommunications and Economic Development. Papers presented at the first International Telecommunication Exposition, Atlanta, Georgia, 9-15 October 1977, pp. 93-98

ULLMAN, J.E.: *Quantitative methods in management* (Schaum's Outline Series, McGraw–Hill, New York, 1976)

VAIL, THEODORE NEWTON: *Views on public questions: a collection of papers and addresses of Theodore Newton Vail, 1907–1917* (Privately printed, 1917)

VICKREY, W.: 'Some objections to marginal cost pricing', *Journal of Political Economy,* **56**, 1948

VICKREY, W.: 'Some implications of marginal cost pricing for public utilities', *American Economic Review,* Supplement, **45**, no. 2, 1955, pp. 605-602

VICKREY, W.: 'Responsive pricing of public utility services', *Bell Journal of Economics and Management Science,* **2**, no. 1, Spring 1971, pp. 337-346

VINOD, H.D.: 'Nonhomogenous production functions and applications to telecommunications', *Bell Journal of Economics and Management Science,* **3**, no. 2, Autumn 1972

VINOD, H.D.: 'Application of new ridge regression methods to a study of Bell System scale economies'. Abstract of paper presented to Telecommunications Policy Research Conference, Airlie, Virginia, April 1975*a*, printed in *Report on Proceedings.* OWEN. BRUCE M. (Ed.)

VINOD, H.D.: 'Bell System production functions, scale economies and public policy'. (Manuscript) Bell Telephone Laboratories, May 1975*b*

WALTERS, A.A.: 'The theory and measurement of private and social cost of highway congestion' *Econometrica,* **29**, 1961, pp. 676-99

WAVERMAN, L.: 'The demand for trunk telephone services in Great Britain'. Paper presented at the Public Economics Workshop, University of Essex, July 1973

WAVERMAN, L.: 'The demand for telephone services in Great Britain, Canada and Sweden'. Paper presented at the International Conference on Telecommunications Economics, University of Aston in Birmingham, 29 May 1974

WAVERMAN, L.: 'The regulation of intercity telecommunications' *in* PHILLIPS, A. (Ed.): *Promoting competition in regulated markets* (The Brookings Institution, Washington DC, Studies in the Regulation of Economic Activity, 1975), chapter 7

WEBB, M.G.: 'The determination of reserve generating capacity critera in electricity supply systems', *Applied Economics,* **9,** no. 1, March 1977

WEINGARTNER, H.M.: 'Criteria for programming investment project selection' *Journal of Industrial Economics,* November 1966

WELCH, S.: *Signalling in telecommunications networks.* (Peter Peregrinus, London, 1978)

WELLENIUS, B.: 'The effect of income and social class on residential telephone demand', *Telecommunication Journal,* **36,** 1969

WENDERS, J.T.: 'An economic framework for the pricing of basic exchange telephone service. Proceedings of the First Symposium on Problems of Regulated Industries, University of Missouri, 1977

WESTFIELD, F.M.: 'Regulation and conspiracy', *American Economic Review,* **55,** June 1965, pp. 424-43

WHITE, C.E.: 'Mobile radio's cautious growth', *Telecommunications,* **11,** no. 8, August 1977

WHITE, C.E.: 'Microwave links are up, up and away', *Telecommunications,* **12,** no. 6, June 1978

WHYTE, A.G.: *Forty years of electrical progress* (Benn, London, 1930)

WILEY, R.E.: 'The US communications consumer and monopoly supply: the FCC position on the proposed legislation', *Telecommunications Policy,* March 1977

WILKES, F.M.: *Capital budgeting techniques* (Wiley, London, 1977)

WILL, T.: *Telecommunications structure and management in the executive branch of government 1900–1970* (Westview Press, Boulder, Co., 1978)

WILLIAMSON, J.: 'F.R. Germany's PTT', *Telecommunications,* **12,** no. 8, August 1978*a*

WILLIAMSON, J.: 'The BPO monopoly: a cause for concern?', *Telecommunications,* **12,** no. 12, December 1978*b*

WILLIAMSON, O.E.: 'Peak load pricing and optimal capacity under indivisibility constraints', *American Economic Review,* **56,** September 1966, pp. 810-27

WILLIAMSON, O.E.: 'Franchise bidding for natural monopolies–in general and with respect to CATV', *Bell Journal of Economics and Management Science,* **7,** no. 1, Spring 1976, pp. 73-104

WILLIG, R.: 'Pricing decisions and the regulator process'. Mimeograph 1979, forthcoming in *Ratemaking problems of regulated industries*

WILLIG, R.D., and BAILEY, ELIZABETH E.: 'Ramsey-optimal pricing of long-distance telephone service' *in* WENDERS, JOHN T. (Ed.): *Pricing in Regulated Industries: theory and application,* Mountain States Telephone & Telegraph Co., Denver, 1977

WISEMAN, J.: 'Uncertainty, costs and collectivist economic planning', *Economica,* May 1953

WISEMAN. J.: 'The theory of public utility price—an empty box', *Oxford Economic Papers,* **9,** 1957, pp. 56-74

WOLMER, RT. HON. VISCOUNT: *Post Office reform* (Ivor Nicholson & Watson, London, 1932)

WOOD, D., and FILDES, R.: *Forecasting for business* (Longman, London and New York, 1976)

WOODS, F.J.: 'Testimony to FCC', Docket 18128, Bell Exhibit 2, 1 April 1970

YATRAKIS, P.G.: 'Determinants of the demand for international telecommunications' *Telecommunication Journal,* **39,** no. 12, December 1972, pp. 732-746

YECHIALI, U.: 'On optimal balking rules and toll charges in the GI/M/1 queueing Process' *Operations Research,* **19,** no. 2, March-April 1971, pp. 349-370

YEH, LEANG P.: 'Satellite communications and terrestrial networks', *Telecommunications,* October 1977

ZAJAC, E.E.: 'A geometric treatment of Averch-Johnson's behavior of the firm model', *American Economic Review,* **60,** March 1970, pp. 117-25

ZAJAC, E.E.: 'Note on 'gold plating' or 'rate base padding', *Bell Journal of Economics and Management Science,* **3,** 1972, pp. 311-5

ZAJAC, E.E.: *Fairness or efficiency: an introduction to public utility pricing* (Ballinger, Cambridge, Mass., 1979)

(b) Subject list

Chapter 1 Introduction

Erlang	(1917)
Gill	(1943)
Grant	(1950)
Hazlewood	(1951)
Kingsbury	(1915)
Kirzner	(1973)
Lee	(1913)
Lewin	(1978)
Mises	(1949)
Morgan	(1976)
Pigou	(1920)
Robbins	(1932)
Shackle	(1972)
Smith	(1776)
Syski	(1960)

Chapter 2 Telecommunications Technology

AT & T	(1975, 1978, annual)
Bell Telephone Laboratory Staff	(1970)
Brooks	(1976)
Brown	(1969)
Clarke	(1958)
Davies and Barber	(1973)
Flood	(1975)
Flowers	(1976)
Haller	(1966)
Hamsher	(1967)
Hills and Evans	(1973)
IEE	(1978)
Joel	(1976)
Jolley	(1968)
Kenward	(1972)
Keyes and Gerla	(1978)
Kingsbury	(1915)
Maddox	(1972)
Martin	(1977)
Merriman	(1975)
Newstead	(1974)
Office of Technology Assessment	(1976)
Phillips	(1969)

Renton	(1974)
Smith	(1974)
Times	(1979)
Welch	(1978)
Williamson	(1978)

Chapter 3 Demand for telecommunications

Alchian and Allen	(1974)
Anderberger and Wikell	(1976)
Auray	(1969, 1970)
Auray	(1978)
Bebee and Gilling	(1976)
Bech	(1967*a,b*)
Betteridge	(1973)
Black and Tryson	(1976)
Bohm	(1970)
Borts	(1964)
Bo *et al.*	(1976)
Brandon and Brandon	(to be published)
CCITT	(1968)
Chaddha and Chitgopeker	(1971)
Cohen	(1976, 1977)
Cracknell	(1977)
Davis *et al.*	(1973)
Deschamps	(1974*a*)
Dobel *et al.*	(1972)
Doherty	(1978)
Drew	(1973)
Drew and Cracknell	(1976)
Dunn *et al.*	(1971)
Evans	(1976)
Ferguson and Gould	(1975)
Flood	(1975)
Florida Public Service Commission	(1975)
Fox	(1973)
Gale	(1976)
General Post Office	(1968)
Gregory	(1973)
Harkness	(1973)
Hayek	(1952)
Hodes	(1963)

Hopley	(1978)
Jipp	(1963)
Johnston	(1963)
Kosberg *et al.*	(1976)
Kotowitz and Waverman	(1973)
Kraepelin	(1958)
Lago	(1970)
Lewis	(1977)
Kipsey	(1963)
Lonnstrom *et al.*	(1967)
Lowry	(1976)
Manus	(1975)
Myskja and Walmann	(1973)
National Board for Prices and Incomes	(1968)
Naoe	(1972)
Nelson	(1969)
Nelson	(1976)
Newstead	(1977)
Perl	(1975)
Pousette	(1976)
Roberts and Ward	(1974)
Select Committee	(1967)
Shackle	(1972)
Simpson	(1969)
Stern	(1965)
Taylor	(n.d.)
Thompson and Tiao	(1971)
Tomasek	(1977)
Troxel	(1947, 1948)
Turner	(1973)
Turner	(1975)
Ullman	(1976)
Waverman	(1973)
Waverman	(1974)
Wellenius	(1969)
Wood and Fildes	(1976)
Yatrakis	(1972)

Chapter 4 Production Functions

Alchian and Allen	(1974)
Cantwell and Honeyman	(1974)
Daniels	(1978)
Davis *et al.*	(1973)
Dobell *et al.*	(1972)
Ellis	(1975, 1977)

Ferguson and Gould	(1975)
Fuss *et al.*	(1979)
Israel, Ministry of Communication	(1974)
Lipsey	(1963)
McElroy	(1975)
Mantell	(1975*a,b*)
Sudit	(1973)
Vinod	(1972, 1975*a.b*)

Chapter 5 Cost functions

Aknai	(1977)
Alchian	(1959, 1968)
Alchian and Allen	(1974)
Alleman	(1977)
Benson and Shurrock	(1974)
Boulter	(1978)
Bowers and Lovejoy	(1965)
Braeutigam	(1974)
British Post Office	(1971)
Clemens	(1950)
Ferguson	(1971)
Ferguson and Gould	(1975)
Froggatt	(1966, 1971*a, b*)
Gabel	(1967)
Kahn	(1970-71)
Lewis	(1949)
Littlechild	(1970*a*)
Littlechild and Rousseau	(1975)
Mandanis	(1976)
Mantell	(1975*a,b*)
Melody	(1971)
Meyer	(1961, 1962)
Mitchell, B.M.	(1978*b*)
Mitchell, D.C.	(1978)
Morgan	(1976)
Rose	(1950)
Simpson	(1945)
Snow	(1975)
Turvey	(1969)
Waverman	(1975)
Yeh	(1977)

Chapter 6 Costa Rica case study

Bonilla	(1977)
Cañas	(1977)
ICE	(1973)
Littlechild	(1973*a*)
Schkolnick	(1976)

Chapter 7 Investment appraisal

Alchian and Allen	(1974)
Biala	(1975)
Bower	(1972)
Chambers and Charnes	(1961)
Charnes *et al.*	(1959)
Cmnd. 3437	(1967)
Cmnd. 7131	(1978)
Collier	(1973)
Gordon	(1962)
Kao and Collier	(1977)
Lonnqvist	(1975)
Lorie and Savage	(1955)
Merrett and Sykes	(1963)
Morgan	(1976)
Reeves	(1977)
Samuels and Wilkes	(1971)
Scherer	(1970)
Simpson	(1977)
Trebing and Howard	(1969)
Weingartner	(1966)
Wilkes	(1977)

Chapter 8 Pricing policy

Alchian and Allen	(1974)
Clemens	(1941)
Ferguson and Gould	(1975)
Hall	(1932)
Hazlewood	(1968)
Jack Faucett Associates	(1972)
Leland and Meyer	(1976)
Lewis	(1949)
Lipsey	(1963)
Littlechild	(1970*a,b,d*, 1975*a,b*)
Littlechild and Rousseau	(1975)
Mitchell, B.M.	(1978*a*)
Naoe	(1972)
Oi	(1971)

Pigou	(1920)
Pyatt	(1972)
Roos *et al.*	(1976)
Squire	(1973)
Troxel	(1947)
Woods	(1970)

Chapter 9 Welfare economics and marginal cost pricing

Alleman	(1978)
Baumol	(1977*a*)
Baumol and Bradford	(1970)
Boiteux	(1949, 1956)
Bonbright	(1961)
British Post Office	(1971)
Carlton	(1975)
Carter	(1977)
Cmnd. 7131	(1978)
Coase	(1946, 1970)
Dreze	(1964)
Faulhaber	(1975)
Feldstein	(1972)
Garfinkel	(1975)
Graaff	(1957)
Hazlewood	(1951)
Hirshleifer *et al.*	(1960)
Hotelling	(1938)
Irvin	(1978)
Kahn and Zielinkski	(1976*a,b*)
Layard	(1972)
Littlechild	(1970*a*, 1975*b*)
Mathewson and Quirin	(1972)
Meade	(1944, 1975)
Meek	(1963)
Mills	(1976)
Millward	(1971)
Mitchell, B.M.	(1978*b*, 1979*a,b*)
Munby	(1968)
National Board for Prices and Incomes	(1968)
Nelson	(1964)
Ng and Weisser	(1974)
Rohlfs	(1979)
Roos *et al.*	(1976)
Roskill Commission	(1971)
Ruggles	(1949)
Select Committee	(1968)
Shepherd	(1966)
Squire and van der Tak	(1975)

Stannard   (1978)
Trebing   (1978)
Turvey (1968*b,c*, 1969, 1970, 1971)
Turvey and Anderson   (1977)
Vickrey   (1948, 1955, 1971)
Wenders   (1977)
Willig and Bailey   (1977)

Chapter 10 Peak load pricing of telephone calls

Baumol   (1972)
Bailey and White   (1974)
Boiteux   (1949)
Craven   (1971)
Deschamps   (1972)
Hazlewood   (1951)
Hirshleifer   (1958)
Hopley   (1978)
Horn   (1977)
Houthakker   (1951)
Johnson   (1966)
Kahn   (1970–71)
Kahn and Zielinkski   (1976*a,b*)
Lewis   (1949)
Littlechild   (1970*c,d/*, 1971)
Littlechild and Rousseau   (1975)
MacGregor   (1978)
Marchand   (1973)
Mitchell, B.M.   (1978*a*)
Mitchell, D.P.   (1978)
de Montbrial and Mossler   (1977)
Muller   (1969)
Pressman   (1970)
Pyatt   (1972)
Roos *et al.*   (1976)
Steiner   (1957, 1971)
Stigler   (1971)
Stigler and Friedland   (1962)
Turvey   (1968*a,b,c*, 1969)
Williamson   (1966)
Willig and Bailey   (1977)

Chapter 11 Grade of service

Aims of industry   (n.d.)
Bishop   (1971)
Boiteux   (1961)
Capello and Sannerio   (1956)

Carlin and Park   (1970)
Cherry   (1958)
Clos and Wilkinson   (1952)
Deschamps   (1974, 75, 1976, 77)
Dreze   (1964)
Jensen   (1950, 57)
Kay   (1968)
Knight, F.H.   (1924)
Knight, N.V.   (1955*a, b*)
Kolm   (1970*a, b*)
Kortanek   (1977)
de Kroes   (1952, 54)
Lave and deSalvo   (1968)
Levy-Lambert   (1968)
Linder   (1969)
Littlechild   (1973*b*, 1974)
Longley   (1970)
Marchand   (1967, 68, 1973)
Mina   (1974)
Morgan   (1976)
Munasinghe   (1977)
Myskia and Walmann   (1972)
Naor   (1969)
Newstead   (1961)
Palm   (1953)
Pigou   (1920)
Pyatt   (1972)
Rapp   (1950, 1961)
Rodenburg   (1949)
Short   (1976)
Smeed   (1964)
Sparrow   (1977)
Syski   (1960)
Tyler   (1973)
Walters   (1961)
Webb   (1977)
World Bank   (1976)
Yechiali   (1971)

Chapter 12 Externalities

American Bell Telephone Co.(annual)
Artle and Averous   (1973)
Borts   (1964)
Dasgupta and Pearce   (1972)
Dickenson   (1977)
Farrell   (1958)
Feldstein   (1972)
Fochler   (1978)
Gellerman   (1977)

Gravelle   (1972)
Hazlewood   (1950)
Horn   (1977)
Irvin   (1978)
Kingsbury   (1915)
Layard   (1972)
Lipsey and Lancaster   (1956)
Little and Mirrlees   (1974)
Little and Scott   (1976)
Littlechild   (1970*a,b*, 1975*b*, 1977)
Lönnstrom and Moo   (1977)
Mayston   (1975)
Millward   (1971)
Morgan   (1976)
Munby   (1968)
Pearce   (1971)
Pyatt   (1972)
Rohlfs   (1974)
Saunders and Warford   (1977)
Squire   (1973)
Squire and van der Tak   (1975)
Turvey   (1968*c*, 1974)
Tyler   (1977)
Vail   (1917)

Chapter 13 Issues in US regulation

Averch and Johnson   (1962)
Bailey   (1973)
Baumol   (1966, 1967,1970, 1971,
                             1977*b*)
Baumol and Klevorick   (1970)
Baumol *et al.*   (1970)
Baumol *et al.*   (1977)
Bolter   (1978)
Braeutigam   (1973*a, b*)
Braeutigam *et al.*   (1974)
Coase   (1959, 62, 65, 66)
Cowan and Waverman   (1971)
Cox   (1971)
Criner   (1977*a, b*, 1978)
Cunningham and Lurin   (1978)
De Butts   (1973, 77)
De Vany *et al.*   (1969)
Dittberner Associates Inc.   (n.d.)
Eads   (1975)
Eastmond   (1970)
Eckert   (1972)
The Economist   (1977)

Ellinghaus   (1967)
Ende   (1969)
Faulhaber   (1975)
The Financial Times   (1979)
Froggatt   (1966, 71)
Gamble   (1978)
Hasselwander   (1978)
Hopley   (1978)
Irwin   (1971)
Johnson, L.   (1961, 1967, 1969,
                       1973, 1976, 1977)
Johnson, N.   (1969)
Kahn   (1968, 71)
Kemp   (1978)
Klass and Shepherd   (1967)
Klein *et al.*   (1971)
Klevorick   (1966, 73)
Kopinski   (1977)
Larkin   (1974)
Lessing   (1967)
Levin   (1962, 66, 68, 70)
Lewis   (1966)
Littlechild and Rousseau   (1975)
Lowry   (1973)
McKie   (1970)
Meckling   (1968)
Melody   (1968, 71)
Morgan   (1923)
Muller   (1979)
NATA   (1976)
Nelson   (1966, 69)
Nicholls   (1963)
Noll   (1971)
Owen   (1973)
Owen and Braeutigam   (1978)
Panzar and Willig   (1977)
Peacock   (1972)
Pearce   (1978*a, b, c, d*)
Peck   (1970)
Phillips   (1969, 74)
Posner   (1969)
President's Task Force   (1968)
Reiger *et al.*   (1963)
Rohlfs   (1979)
Sandberg   (1975)
Scherer   (1970)
Sheahan   (1951, 56)
Sharkey and Telser   (1977)
Shepherd   (1966, 70, 71, 74)
Thayer   (1966)
Trebing   (1963, 68, 69, 71, 74)

Trebing and Howard (1969)
Trebing and Melody (1967)
Troxel (1966)
Waverman (1975)
Westfield (1965)
White (1977, 78)
Wiley (1977)
Will (1978)
Willig (1979)
Zajac (1970, 72)

Chapter 14 Some alternative views

Alchian (1965, 69)
Alchian and Allen (1974)
Alchian and Kessel (1962)
AT & T Annual Report (annual)
Baer and Mitchell (1975)
Baldwin (1925)
Bennett (1895)
Billingsley (1973)
Buchanan (1969, 68, 1975)
Buchanan and Thirlby (1973)
Buchanan and Tollison (1972)
Buchanan and Tullock (1962)
Canes (1966)
Chapuis (1978)
Cmnd. 3437 (1967)
Cmnd. 7131 (1978)
Crutchley (1938)
De Alessi (1974)
Demsetz (1968)
Elrom (1972)
Furubotn (1972)
Gabel (1969, 74)
Gradnmont (1977)
Hahn (1973)
Hai (1976)
Hayek (1937, 48)
Holcombe (1911)
Jordan (1972)
Kingsbury (1915)
Kirzner (1973, 78)
Knight (1924)
Kohlmeier (1969)
Lee (1913)
Littlechild (1976, 1978)
Lorenz (1978)
Maddox (1972)
Menger (1871)
Meyer (1907)
Miller (1975)
Mises (1949)
Murray (1927)
Owen (1978)
Pearce (1978a – e)
Peltzman (1971)
Phillips (1975)
Poole (1979)
Posner (1971, 77)
President's Task Force (1968)
Pye (1977)
Reuss (1975)
Robertson (1947)
Roos et al. (1976)
Shackle (1961, 72)
Sheahan (1951)
Stigler (1971, 61)
Stigler and Friedland (1962)
Stone (1973)
Thirlby (1960, 73)
Waverman (1975)
Whyte (1930)
Williamson (1976, 78)
Wiseman (1953, 57)
Wolmer (1932)